COLORADO GEM TRAILS

AND MINERAL GUIDE

RICHARD M. PEARL

★

COLORADO
GEM TRAILS
AND
MINERAL
GUIDE

★

SKETCH MAPS BY MIGNON WARDELL PEARL

THIRD, REVISED EDITION

SWALLOW PRESS
OHIO UNIVERSITY PRESS

ATHENS

Dedicated to

JIM AND JO LITTON

NATIONAL FORESTS IN COLORADO

Name	*Headquarters*
Arapaho	Golden
Grand Mesa	Delta
Gunnison	Gunnison
Pike	Colorado Springs
Rio Grande	Monte Vista
Roosevelt	Fort Collins
Routt	Steamboat Springs
San Isabel	Pueblo
San Juan	Durango
Uncompahgre	Delta
White River	Glenwood Springs

NATIONAL MONUMENTS IN COLORADO

Black Canyon of the Gunnison National Monument	Montrose County
Colorado National Monument	Mesa County
Dinosaur National Monument	Moffat County
Florissant Fossil Beds National Monument	Teller County
Great Sand Dunes National Monument	Saguache and Alamosa Counties
Hovenweep National Monument	Montezuma County
Yucca House National Monument	Montezuma County

NATIONAL PARKS IN COLORADO

Mesa Verde National Park	Montezuma County
Rocky Mountain National Park	Grand and Larimer Counties

4

The following magazines are national publications devoted to mineral and gem collecting on a popular level. They report new finds and new facts.

Gems and Minerals, P.O. Box 687, Mentone, California 92359. $4.50 per year. Don MacLachlan, editor.

Rocks and Minerals, P.O. Box 29, Peekskill, New York 10566. $4.00 per year. James and Winifred Bourne, editors.

The Lapidary Journal, P.O. Box 2369, San Diego, California 92112. $5.75 per year.

Earth Science, P.O. Box 550, Downers Grove 60515. $3.00 per year. Mary G. Cornwell, editor.

BOOKS ON MINERAL COLLECTING

How to Know the Minerals and Rocks, by Richard M. Pearl, published in 1955 (and paperbound in 1963) by McGraw-Hill Book Company, New York, and in a Signet Science (paperbound) edition by the New American Library of World Literature, Inc., New York. A practical field guide to more than 125 important minerals and rocks, featuring basic keys for identifying typical specimens without special skill or equipment; includes many marked drawings and color plates.

Minerals of Colorado: A 100-Year Record, by Edwin R. Eckel, published in 1961 as U. S. Geological Survey Bulletin 1114. A summary of the known facts about all the Colorado minerals.

Exploring Rocks, Minerals, Fossils in Colorado, by Richard M. Pearl, revised edition published in 1969 by Sage Books (The Swallow Press, Inc., Chicago). A well-illustrated survey of Colorado's geology and collecting localities for the earth-science student and hobbyist.

Getting Acquainted with Minerals, by George Letchworth English and David E. Jensen, 2d edition published in 1958 by McGraw-Hill Book Company, New York. A well-illustrated, introductory book, which includes discussions of mineral properties, descriptions of minerals, and information for collectors.

American Gem Trails, by Richard M. Pearl, published in 1964 by McGraw-Hill Book Company, New York. A survey of gem localities in which Colorado and adjoining states are well represented.

Minerals and How to Study Them, by Edward S. Dana and Cornelius S. Hurlbut, Jr., 3d edition published in 1949 by John Wiley & Sons, Inc., New York. An elementary book on mineralogy, also available paperbound.

1001 Questions Answered About the Mineral Kingdom, by Richard M. Pearl, published in 1949 by Dodd, Mead and Company, New York, and in 1960 (paperbound) by Grosset and Dunlap, New York. Answers the questions most usually asked about all aspects of minerals and collecting.

PREFACE

Colorado Gem Trails and Mineral Guide takes you on an extensive collecting tour of a great mineral-producing state. It directs you to the best localities, whether world famous or little known, revealing the places where Nature has hidden her specimens of mineral wealth and beauty. This book is intended to complement my *Exploring Rocks, Minerals, Fossils in Colorado*, also published by Sage Books (The Swallow Press, Inc., Chicago, revised edition, 1969).

The localities represent every section of Colorado, especially the central part of the state and the Denver area. Each locality has been visited and examined by the author and his wife within recent weeks. Aided by mileage logs, sketch maps, and highway maps, detailed instructions are given for reaching all the localities described. The latest information is presented on land ownership and any local condition affecting transportation, accessibility, or collecting. Directories of mineral societies and museums, references for further study, and lists of official maps are given. An original system of indicating routes and places is an important and very useful part of this book. (Pages 16 to 19 should be read for a further explanation of these special features.)

Under the title *Colorado Gem Trails*, the predecessor of this book was published by Sage Books, Denver, in 1951. Two paperback reprints, referred to as second and third editions, were then issued by Mineral Book Company, Colorado Springs and Denver. The present title was first published by Sage Books, Denver, in 1958; a second, revised edition was issued in 1965; it is now presented as a third, extensively revised edition, but if the original title is included, this is the fourth edition, tenth printing.

In addition to the many persons whose names are mentioned in the text, the author wishes to acknowledge personal assistance from the following friends:

John H. Alexander, Timothy Anglund, W. Russell Bailey, John S. Baird, Clark F. Barb, Alfred E. Birdsey, F. Martin Brown, Mrs. Pauline L. Bryan, J. E. Byron, W. B. Carstarphen, Vernon H. Cato,

Robert L. Chadbourne, Maurice Christensen, Major Roy Coffin, Clarence G. Coil, Muriel Colburn, James H. Cosgrove, Dr. Ralph D. Crawford, Dr. H. C. Dake, H. W. Endner, Max Fillmore, Hazel and Richard Fischer, Dr. William F. Foshag, Dr. Russell D. George, Cecil F. Goldsworthy, Dr. Don B. Gould, J. E. Graf, S. N. Green, Dr. Clifford C. Gregg, Thomas Haegler, Al J. Hall, Mrs. Liane Hall, Mrs. Mildred W. Hawthorne, Harold and Vera Hofer, Virginia Holbert, Chester R. Howard, Raymond H. Huck, Jerome Hurianek, James R. Hurlbut, Mrs. Robert Hymen, Virginia Irvine, Estle Johnson, Keith Johnson, Frank C. Kessler, Parnell King, Dr. Kenneth K. Landes, Dr. John H. Lewis, Prof. Burton O. Longyear, Harvey C. Markman, Dr. Ellsworth Mason, Carl F. Mathews, Prof. Arthur J. McNair, Fred McNair, Will C. Minor, Frank Morse, Jessie Myers, H. W. Nichols, Harry W. Oborne, Donald J. Olson, Edwin W. Over, Jr., Mrs. Andrew Pennoyer, George Robertson, Mrs. Lottie M. Shipley, G. Ruegg, P. J. Schlosser, Col. George A. Sense, Col. Calvin B. Simmons, Dr. George Switzer, Ed Tezak, Jr., Ron A. Timroth, R. C. Vance, Stephen Varni, George M. White, Herbert W. Whitlock, Robert D. Wilfley, Frank B. Wilson, Mrs. Ruth Wright, Raymond F. Ziegler.

Photographs have been furnished by Clarence G. Coil, Dale Hollingsworth, James R. Hurlbut, David E. Jensen, and George Veto.

8

CONTENTS AND MAPS

9

MINERAL COLLECTORS' PARADISE

Colorado – the Centennial or Silver State – is more noted for the variety of its minerals than for the size of its mineral deposits. Only two of its many mines – the Climax molybdenum mine at Climax, near Leadville, and the Eagle zinc mine at Gilman, with its unique underground mill – can be regarded as really large operations. No single one of Colorado's individual mines, moreover, presents the incredibly large number of minerals that is so astounding a feature of such deposits as those of the New Jersey Zinc Company at Franklin and Sterling Hill, New Jersey, or the Crestmore quarry at Riverside, California.

Nevertheless, Colorado ranks as one of the most strongly mineralized areas in the world. In the sixth edition of *Dana's System of Mineralogy*, published in 1892, it led all the rest of the states in the number of different minerals listed. Since then, it has been surpassed by the larger state of California, but the variety of Colorado minerals is considerably greater than in most of the other states, as enumerated by Edwin B. Eckel (Ref. 1), who listed 445 species up to 1957.

This exceptional diversity, without undue concentration in just a few places, means that Colorado minerals are widely disseminated in numerous small properties, which makes them more readily accessible to the collector. Another favorable aspect is that so many of Colorado's prolific mineral localities are on public land, available to all collectors on equal terms. This contrasts with the much more limited opportunities in certain other parts of the country.

The enormous mineral production that has come from Colorado since 1858 is therefore largely the output of many small deposits combining to yield an aggregate of 3 billion dollars' worth of metals, in addition to another 3 billion dollars in nonmetals and fuel resources. Colorado stands high in the total American production of gold, silver, copper, lead, zinc, uranium, radium, vanadium, and molybdenum. Minerals of all these metals and many

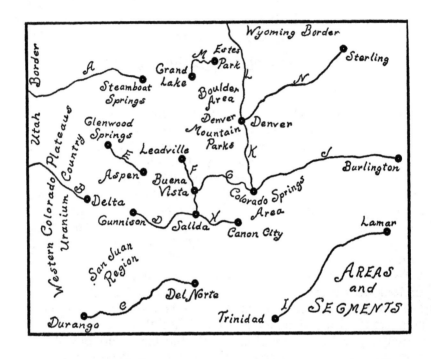

more await the collector in Colorado's mountains, plateaus, and plains.

Ore minerals are by no means the only kind found in the state. At least 30 species and major varieties of gems, besides the organic substances amber and jet, are known in Colorado. Many other useful nonmetallic, or so-called industrial, minerals occur here also, as well as a large number of minerals that do not yield anything of value but that are inherent parts of the rocks that constitute the earth's crust: these are called rock-forming minerals. Still other minerals, not included in the above categories, make their home in stream gravels and rock cavities, and they also may be sought successfully throughout this mineral-rich state.

As might be expected, many minerals were found first in Colorado. More than 40 of them have been named after Colorado persons or places. These places include mines, streams, mountains, towns, counties, and the state itself. Coloradoite, a telluride of mercury originally found in the Keystone mine, Magnolia mining district, Boulder County, was the first mineral to bear a Colorado name. Brief accounts of the discovery and naming of these minerals have been given in a series of articles by the author of this book (Ref. 2-6). Localities where some of these minerals can still be obtained from their historic sources are included in this book.

The experienced field collector of minerals recognizes the value of considering the geology and topography of the region in which he is working. Thus, he would expect to find the sulfide and other ore minerals in the mountainous parts of Colorado, because ore veins are so generally affiliated with mountain-making processes. Weathered specimens of minerals such as pyrite, chalcopyrite, galena, and sphalerite can be picked up on almost any mine dump in the hills. With them will be seen the usual gangue minerals, including quartz, barite, rhodochrosite, and fluorite. The splendid array of pegmatite minerals would likewise be expected in the "hard rock" of the mountains.

The veteran collector would look for nodules and aggregates of certain common minerals in the sedimentary formations flanking the mountains and spreading across the plains. Here are found the interesting twins of aragonite, groups of marcasite, and cubic pseudomorphs of goethite after pyrite. Here, too, and in the sedimentary rocks of the plateaus and the flat areas called parks, occur the agate, jasper, and petrified wood in which Colorado abounds.

This bird's-eye method of collecting has many practical limitations. Yet, a good deal about the possibilities of a given part of the state can be learned by careful examination of a geologic map, or even a topographic map, of Colorado. The mineral maps of the state show clearly the strategic situation of the most densely metallized area, which stretches in a southwest direction in a broad belt from Jamestown in Boulder County to the La Plata Mountains in the "silvery San Juans."

One of the chief reasons for the astonishingly long catalog of different minerals in Colorado is the variety of its geology, which in turn is attractively reflected in its topography. Bounded by parallels of latitude and meridians of longitude, which cut at random across the physical features of the landscape, Colorado partakes of the character of each of the seven adjacent states that touch its borders. Only Tennessee and Missouri, with eight, have a larger number of neighbors. The Four Corners in southwest Colorado is the only place in the nation where four states come together. Furthermore, Colorado lies within five natural regions or physiographic provinces, it being exceeded by only one other state, namely Alaska.

Wide north-south zones mark the three main regions of Colorado. On the west, entering the state from Utah, is the Colorado Plateau, ranging from 5,000 to 11,000 feet above sea level. The central zone is the Southern Rocky Mountains, which culminate in the highest 14,000-foot peaks of the Sawatch Range, and which include intervening valleys and open spaces known as parks. On the east are the Great Plains, descending from an altitude of 6,500 feet at the foothills of the mountains, and reaching a minimum height of 3,350 feet where they move into Nebraska and Kansas. Two smaller natural regions are represented by the southeastern edge of the Middle Rocky Mountains, separated from the other Rockies of Colorado; and the Wyoming Basin, a rolling prairie broken by streams and isolated mountains.

Each of these five areas is distinguished by its own type of geology, its own kinds of rocks, and its own assortment of minerals. Certain ubiquitous minerals are, of course, not restricted to specific areas. Gold, for example, is "where you find it," but even gold is prospected for with better results in some rocks than in others. The "lucky" mineral collector is often the one who has a keen sense of the rock relationships and mineral associations in the place where he is working, based usually on extensive experience.

A large number of choice localities are described in this book. Each has been visited and examined by the author within recent weeks. These localities have been selected for inclusion according to the following criteria:

1. Emphasis is given to places where collecting for worthwhile specimens can be done with reasonable expectation of success. Owing to the phenomenal increase in the number of mineral collectors who pursue their hobby in Colorado, and the continuing depletion of many of the available localities, some are described that fall short of being outstanding ones but still seem worth a visit. Others are merely mentioned in passing so that the collector may know about them while he is in the vicinity. Furthermore,

16

this book has been written for the student as well as the collector, and some of the localities may not be remarkably prolific, although all are desirable. The former editions of *Colorado Gem Trails*, however, purposely included a number of gem deposits that were once productive but are now of historical interest only.

2. The localities should be at least fairly accessible by regular automobile; Mount Antero has been the chief exception, and difficult roads are clearly designated.

3. Preference is given to localities on public land or where no objection has hitherto been made to individuals collecting specimens for their personal use, as opposed to commercial collecting. The petrified forests require a modest admittance fee. Other good localities described may at the moment be closed, but it is to be hoped that the situation will change. The author and publisher must obviously decline responsibility for the activities of others, but suggest to the reader caution and consideration at all times where property rights and public laws are involved. Property ownership and mining rights may, of course, change hands at any time, and so the collector needs to be aware of the necessity for securing permission before trespassing upon land that is fenced, posted, or otherwise restricted.

Numerous mileage logs, believed to be the most useful kind of directional information, are incorporated; never before published, these have been prepared solely for this book. It should be remembered, however, that automobile speedometers vary in recording distance – they are nearly always in excess – and so the figures should be checked as one proceeds. The mileages given are corrected ones. Sketch maps made exclusively for this book by Mignon Wardell Pearl provide an overall view of large areas and furnish a helpful guide to difficult localities.

Special conditions affecting transportation, accessibility, or collecting in a particular place are given whenever known to the author. Land ownership has been investigated insofar as possible at the appropriate county, state, or federal agency.

General instructions, precautions, and practical advice on clothing to wear, equipment to use, and procedures to follow in collecting minerals on field trips are not included here but can be read in the books on mineral collecting listed on pages 9-10. Many of these ideas pertain to any outdoor activities, but a few have personal meaning only for mineral collectors and should be care-

17

fully digested. The regulations in effect in the national forests are printed in the Forest Service maps listed for each locality. New rules are being issued regarding the amount of petrified wood that can be collected within the forests, and it may be desirable to inquire about them before taking away any large quantity of material. The cleaning, care, and display of crystals and mineral specimens after they are found is discussed in *Cleaning and Preserving Minerals*, by Richard M. Pearl.

Mineral specimens found in Colorado will be identified free to residents ($1.00 per sample for out-of-state residents) if sent to Prospector Mineral Identification Service, P.O. Box 112, Golden, Colorado 80401.

Museums that contain worthy mineral collections open to the public, and mineral societies that welcome visiting collectors to their meetings are also mentioned here. The museums provide the collector with an opportunity to see good specimens of local material, and the societies enable the traveler in the field to become acquainted with residents of like interests, some of whom may be able to furnish an intimate knowledge of nearby localities.

The references noted are chiefly those that contain significant basic information, but there are a somewhat longer number of entries than were employed in the earlier edition of this book. Full credit is given throughout to the individual collectors and scientists whose diligent efforts have made this work possible.

Free or inexpensive government maps covering specific areas are listed in this book. They are designated as either topographic or national-forest maps. The topographic maps are sold, in person or by mail, by the U. S. Geological Survey, Federal Center, Denver, Colorado 80225. They are also sold over the counter in the downtown office of the Federal Building, in Denver. The maps belonging to the series of standard quadrangles cost 50 cents each; some can also be bought at local book and stationery stores at slightly advanced prices. The topographic maps of special areas are sold at the various prices quoted in this book. To keep abreast of the completed federal topographic mapping projects in this state, the reader should request the current Index to Topographic Maps of Colorado, obtainable free. The U. S. Forest Service maps are distributed without charge at the offices in Washington, D. C., at the Federal Center (Denver) and in many of the cities and towns in Colorado. Owing to irregularities in the forest boundaries, these maps overlap, so that many of the localities appear on more than one map, all of which are listed alphabetically under the heading "Maps." The Colorado Department of Highways, 4201 East Arkansas Avenue, Denver 80222, sells detailed county maps at 75 cents; most counties require more than one map. Larger maps, printed to order, sell for $3.00. Free state, regional, county, and local maps are available from gasoline companies, chambers of commerce, and the Colorado Department of Highways.

The method of denoting the routes and localities in *Colorado Gem Trails and Mineral Guide* has been devised expressly for this book. The main highways are divided into segments, from which diverge the roads leading to the individual localities. A letter indicates the major segment of the main highway near which the locality is situated. Thus, for example, as shown in the table of contents, the localities under "D" may all be reached from U. S. Highway 50 between Gunnison and Salida, while those under "H" are accessible from the same highway between Salida and Canon City. Many of the localities may, of course, be reached on other roads from other directions. The larger towns, with their superior facilities, have been named as the starting and terminal junctions between the chief segments of the tours. Five general areas are described with logs but without highway segments.

REFERENCES

1. U. S. Geological Survey Bulletin 1114, 1961.
2. Colorado Magazine, vol. 18, 1941, p. 48-53.
3. Colorado Magazine, vol. 18, 1941, p. 137-142.
4. Mineralogist, vol. 16, 1948, p. 59-61.
5. Mineralogist, vol. 19, 1951, p. 283-286.
6. Mineralogist, vol. 20, 1952, p. 70-72.

HISTORY OF COLORADO GEMS

Almost from the earliest days of its gold and silver discoveries, Colorado has held a leading place in the production of gems, ranking among the first half-dozen states in value of output. Nonmetallic minerals in general were, however, given attention later than the metallic ones and were not exploited so soon.

There has long been, of course, more or less isolated finds of gemstones that received the benefit of local gossip but were rarely made known to the outside world. Other than these, the first gems were obtained by scientists who used them for purposes of study and publicized them widely. Then prospectors, individually or as small companies, mined them commercially. In recent years, mineral collectors and rockhounds, including part-time dealers, have become responsible for an increasingly large proportion of the discoveries.

But the history of the gems of Colorado goes back far beyond hobbyist or prospector or scientist. It is to the building of the Southern Rocky Mountains, which occupy the central of the three major north-south topographic divisions of the state, that Colorado is indebted for a large proportion of its rich mineral resources, and the gems are no exception. Most of them either were formed by processes associated with the mountain-making activities or, having originated earlier, were exposed by the uplift (and subsequent erosion) that followed most of the compressional folding. As advertised by Pohndorf's, the former Denver mineral dealer, these gems are true Western antiquities!

The infamous Great Diamond Hoax of Western lore took place within the boundaries of Colorado. Although it scarcely pertains to the natural history of the state, it does bear a synthetic relationship to the history of Colorado gems. It was the most expensive land fraud in the early days of the West.

The first list of Colorado gems to appear formally seems to have been that included in the classification of metals and minerals prepared by J. Alden Smith of Orvando J. Hollister's fascinating book *The Mines of Colorado*, published in 1867 by Samuel Bowles and Company, of Springfield, Massachusetts.

Successive lists by Smith and other writers are enumerated in *Exploring Rocks, Minerals, Fossils in Colorado*, by Richard M. Pearl (Sage Books, The Swallow Press, Inc., Chicago, revised edition, 1969) and *Minerals of Colorado: A 100-Year Record*, by Edwin B. Eckel (U. S. Geological Survey Bulletin 1114, 1961).

There are a few men who were scientists or analysts and whose names the reader encounters repeatedly in the gem literature of Colorado's earlier years: R. T. Cross, Samuel L. Penfield, Walter B. Smith, Whitman Cross, W. F. Hillebrand, George F. Kunz, and, more recently, Douglas B. Sterrett, the last two of whom wrote the annual chapters on precious stones in *Mineral Resources* for the U. S. Bureau of Mines.

Many were the prospectors who searched the mountains of the state for signs of "color," for surface crystals of gems, and for the structures that might be likely to yield more of them. Quartz veins and pegmatite dikes were regarded as especially indicative of gem minerals. Several of the men were more generously endowed than the rest with the qualifications that make for successful mineral hunting, as well as with a bit of luck, and they were able to report rather frequent discoveries over a number of years.

No name appears more often than that of J. D. Endicott, of Canon City. During the beginning decade or so of the present century, especially around 1908, he found, claimed, and worked a variety of gem deposits throughout central Colorado.

The famous gem region around Mount Antero, in the Sawatch Range, must always be associated with Nathaniel D. Wanemaker, who perhaps made the first discovery of aquamarine there in 1884 or 1885. He lived for years in a small stone cabin in the glacier-gouged amphitheater on the south side of the mountain, about 800 feet below the summit. The roofless ruins of the old cabin still stand, entirely surrounded by barren rock, the only timber 1,500 feet below, the only water from a small pond that dries up before the end of summer.

W. C. Hart, who came to Manitou Springs in 1892, was another early veteran in search of gems in Colorado, and his activities extended almost to the Wyoming border. He issued a little book of localities and assembled the collection that was displayed at the World's Columbian Exposition in 1893 and is now in the Pioneers' Museum in Colorado Springs. After his death in 1936 at the age of 85, Mr. Hart's two daughters continued for a while to operate his Rocky Mountain Gem Store in Manitou Springs, where part of his collection could be seen.

Albert B. Whitmore, who came to the Pikes Peak region in 1874 and mined at Cripple Creek, where he also maintained a specimen store, was one of the more interesting characters of those years. He homesteaded at Crystal Park, where he lived almost as a hermit for over 35 years, although he was cordial to visitors who came to look at his stock of specimens, often staying to buy. He operated the Crystal Peak Gem Company in the area that became known as the Gem Mines and then the Crystal Gem Mines.

Rev. R. T. Cross, a resident of New York state, arrived in Colorado Springs in 1876, becoming pastor of the Congregational Church. He visited the important localities that were then just beginning to produce, first identified the gem zircons at St. Peters Dome, and was instrumental in calling the attention of scientists to the significant discoveries of other collectors. Two of Mr. Cross' books, *Clear as Crystals* and *Crystals and Gold*, indicate his interests; the latter book is an intriguing story about his collecting experiences in Colorado and elsewhere.

Edwin W. Over, Jr., who was a resident of Woodland Park when he died in 1963, was excelled by no other professional collector in his skill at finding specimens. His chief work in Colorado was in the Pikes Peak region, and many of his best specimens are to be seen in the large eastern museums.

The history of the activities of other mineral collectors and dealers of the Pikes Peak region has been carefully compiled by George M. White, of Colorado Springs, though not yet published, except the part dealing with Crystal Peak (Ref. 1).

In a vote conducted in 1949 by the Colorado Mineral Society, in which other collecting groups were invited to participate, aquamarine was selected as the first choice for the state gem of Colorado. The author's personal vote for the first three gems would be for amazonstone, turquoise, and aquamarine. The last one was named the official state gem in 1971.

The most characteristically Colorado gem is amazonstone. It is a variety of microcline feldspar, which in its usual form is a common and widespread mineral. Amazonstone occurs in bright green and blue, and blends of the two colors, often diluted with gray. It is opaque, and so its appeal is entirely one of color and luster. Often mistaken for jade, it is frequently sold as "Colorado jade" or "Pikes Peak jade." The first notice of amazonstone seems to have been in 1867 by Ovando J. Hollister, who mentioned it as occurring with several other minerals at the head of Elk Creek, "five miles from the old St. Louis Ranch." A. C. Peale of the Hayden Survey wrote in 1873: "About the base of the peak [Pikes] I found, rather abundantly, good crystals of amazonstone (green feldspar) and smoky quartz." Colorado amazonstone was made known to the world by a large display of it at the Centennial Exposition in Philadelphia in 1876, and the quality and quantity of the specimens and their low prices drove the Russian material from the market and brought grief to the exhibitors who had shipped much of it from Europe for sale at the fair. The Pikes Peak region is still the most important source of amazonstone, and sales of cut gems, mostly to tourists in Denver and Colorado Springs, went above $1,000 annually during a number of past years. Present-day localities for amazonstone are described at appropriate places later in this book.

Colorado ranks among the leading states in the production of turquoise. The first published mention of this gem in Colorado appears to have been in 1870 by J. Alden Smith, who had in his collection specimens cut in keystone form, drilled and formerly worn in a bracelet, which he had obtained from a Ute chief. The stones were supposed to have come from an uncertain locality in southern Colorado. "They are highly prized by the Indians," he said, "and it is with much difficulty that they can be induced to part with them." This is a good epitome of the whole history of turquoise in Colorado: its basis, the worship by the Indians; its use, in rude but interesting ornamental and talismanic jewelry; and its source, mostly in the southern part of the state, but the precise places formerly obscured by mystery and legend. Several turquoise localities in Colorado are mentioned later in this book.

Although aquamarine is found elsewhere in the state, the one place that accounts for its importance as a leading Colorado gem is Mount Antero, high in the Sawatch Range. Specimen isolation at a lofty altitude, the interesting mineralogy of the area, the peculiar

23

characteristics of the crystals, and the beauty of the material when fashioned have all combined to make Antero minerals of exceptional appeal to collectors, dealers, and students.

In addition to these three gems, Colorado is an important source of others. Its topaz is second to none in the United States in quality. Its smoky quartz is superb, especially as crystal specimens when associated with amazonstone. Colorado is one of the few places in the world where first-rate lapis lazuli occurs, and one of the few in the country where gem sapphire has been produced. It is a major supplier of alabaster and has been a world source of jet. Large amounts of the chalcedony-quartz gems — agate, jasper, and special varieties, such as petrified wood and agatized dinosaur bone — are found in a wide diversity of colors and patterns. Amethyst, rock crystal, rose quartz, garnet, and obsidian are among the other worthwhile Colorado gems described in this book.

The excellent quality of many of Colorado's gem minerals has made them widely known. The material from the Pikes Peak region, even when found up to 40 miles from the mountain itself, has more often than not been labeled merely Pikes Peak, especially the stones that have been sold in tourist jewelry, because of the attraction and significance of the name. Much of the supply from elsewhere in the state, however, has been sold without any indication as to its source. Turquoise, for an example, is second or third among the gems of the United States in total value of production to date, yet encouragement of public appreciation of Colorado turquoise has been greatly neglected.

The future of the gem industry of Colorado seems bright. Both amateur and professional mineral prospectors, as well as specimen collectors, are increasing in number. Tourists to the West can usually be counted upon to buy gem-set jewelry for souvenirs and gifts. Their further acquaintance with Colorado gems, and especially the existence of a growing group of lapidaries and the increased interest everywhere in minerals as a hobby, will probably have the effect of expanding Colorado's gem production in the years to come. When the high value of gemstones is considered in relation to their bulk, it becomes rather obvious that only a very small area is required to yield enough to surpass previous finds in importance. In general, the more rugged the topography, the more highly mineralized is the crust of the earth and the less thoroughly explored is the land. The search for rare substances of industrial use may be the most successful means of disclosing the great gem mines of the future; emeralds, for instance, have been found (though not yet in Colorado) while investigating deposits of beryllium, one of the elements of which they are composed. It is the author's hope that readers of this book may be among those who will find the best gems!

REFERENCES

1. Rocks and Minerals, vol. 10, 1935, p. 184-187.
2. *American Gem Trails*, Richard M. Pearl, McGraw-Hill Book Co., 1964.
3. U. S. Geological Survey Bulletin 1114, 1961.

WESTERN COLORADO

The wonderfully varied strip of western Colorado, facing the state of Utah, has come into its own as an incalculably rich source of mineral wealth. The San Juan Mountains, in the southern part of the region, have long been noted for their yield of precious and base metals. With the discovery, moreover, of oil and gas in the San Juan Basin, southwest of the mountains, and the production of oil from greater depths than before at Rangely, well to the north, together with the feverish activity that occurred in the uranium deposits between those places during the 1950's, the search for mineral wealth stretched from New Mexico to Wyoming.

Part of this western zone of Colorado is included within the San Juan Mountains. The vicinity of Dinosaur National Monument, also discussed later, is regarded as part of the Middle Rocky Mountains, separate in topography and orientation from the other Rockies of Colorado. North of the monument area is an open section of Moffat County belonging to the Wyoming Basin. The rest of this western zone is in the Colorado Plateau province, of which the so-called uranium country, described later, is a considerable part.

MAPS

The Index to Topographic Maps of Colorado, obtainable free from the U. S. Geological Survey, Federal Center, Denver, Colorado 80225, shows which parts of western Colorado are covered by topographic maps. The national-forest maps of this region are Grand Mesa, Gunnison, Routt, San Juan, Uncompahgre, and White River.

UTAH BORDER TO STEAMBOAT SPRINGS

Traversing rolling pasture land and range preserve, U.S. 40 enters Colorado from Vernal, Utah, crossing the Uinta Basin and skirting Dinosaur National Monument (described below) as it continues across Moffat and Routt Counties, on to Steamboat Springs. The silica gems of Moffat County and the Hahns Peak locality are described below. The oil town of Rangely is only a short distance south of this highway, and other oil-bearing structures are not far away. The uranium country (described later) stretches far to the south. The hot springs indicate other geologic activity in this vast expanse of northwestern Colorado — at once desolate and fascinating.

Moffat County

The silica specimens of Moffat County are typical of western, dry-land material: jasper of red, brown, yellow, and green colors;

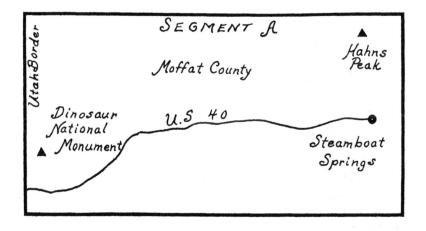

agate and agatized wood; opalized wood; and petrified and agatized dinosaur bone.

The distribution of the silica gems in Moffat County has been detailed by Clark F. Barb (Ref. 1), who notes that most of the agate, jasper, chert, flint, and opal in this part of the United States (including the corners of Wyoming and Utah) comes from the Morrison formation (Jurassic age), Chinle formation (Triassic age), and Morgan formation (Pennsylvanian age). Both the outcrops and the sedimentary material eroded from them yield large amounts of agate. The presence of slabs of agatized wood are very diagnostic of the conglomerate at the base of the Fort Union formation. The best localities are those containing outwash material from the south and east flanks of the Uinta Mountains, especially the gravel capping the hills along U.S. 40 and north, between the Utah border and Cross Mountain (which is 15 miles west of Maybell). The areas between Cross Mountain and Craig, and from Craig north, are relatively barren, as is the area south of U.S. 40.

Professor Barb has described numerous specific localities, telling how to find them, although not all can be reached in an ordinary automobile. Anyone having several days to spend in Moffat County can put his careful instructions to good use.

Theodore H. Kleeman (Ref. 2) reported that petrified wood is found along U.S. 40, being especially abundant near Craig.

L. Ray Clemmer (Ref. 3) mentioned an abundance of good agate pebbles 25 miles east of Artesia and then 1 mile north between Little Wolf and Big Wolf Creeks until reaching a hill on the left.

REFERENCES

1. Mineralogist, vol. 26, 1958, p. 147-151, 198-201.
2. Mineralogist, vol. 9, 1941, p. 253.
3. Gems and Minerals, no. 333, June 1965, p. 60.

MAPS

The Index to Topographic Maps of Colorado, obtainable free from the U. S. Geological Survey, Federal Center, Denver, Colorado 80225, shows which parts of this area are covered by topographic maps.

Dinosaur National Monument

Although collecting is prohibited within Dinosaur National Monument itself, petrified dinosaur bone, agate, and jasper similar to those found over a wide part of northwestern Colorado occur inside the monument. The dinosaur remains are in the Morrison formation, of Jurassic age, and the petrified wood and other materials are in the Dakota formation, of Cretaceous age. Attractive amethyst quartz occurs in Dakota chert.

The presence of dinosaur bones embedded in the rocks of the Uinta Mountains was known at least by 1882, but the remarkable concentration at the present site of the quarry was discovered by Earl Douglas in 1909. Between 1909 and 1922, about 700,000 pounds of bone and rock was shipped to the Carnegie Museum, at Pittsburgh, which operated the quarry. The most remarkable fossil skeleton obtained from this locality was of a *Brontosaurus*, 71½ feet long and 15 feet high, weighing when alive about 35 tons. *Diplodocus* was even longer, reaching 75½ feet in length. The most abundant of the dinosaurs here was *Stegosaurus*, the armor-plated one that had "two sets of brains." These dinosaur finds are discussed in *Exploring Rocks, Minerals, Fossils in Colorado*, by Richard M. Pearl (Sage Books, The Swallow Press, Inc., Chicago, revised edition 1969).

This region is situated within that small part of the Middle Rocky Mountains that lies in Colorado, near the east end of the Uinta Range. These mountains run eastward almost perpendicularly to the rest of the Rockies, and they constitute perhaps the largest range having that direction on the American continents. Dinosaur National Monument extends across the border from Moffat County into Uintah County, Utah, and is only a few miles south of Wyoming. It covers a quarter of a million acres of primitive wilderness, as wild a western country as can be found. In this area is the junction of the Green and Yampa Rivers, which flow together through majestic gorges surpassed by few other canyons anywhere in the world. Lodore Canyon in places is more than 3,000 feet deep.

The main entrance to the monument is from Jensen, Utah, on U.S. 40. The Dinosaur Quarry and Visitor Center, including a unique museum where you can watch dinosaurs being brought to light for the first time since their burial, are 7 miles northeast of

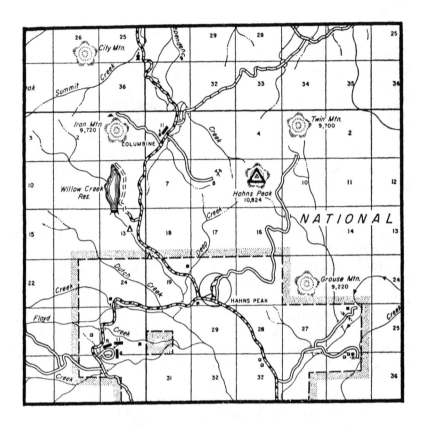

Jensen. The headquarters is at Dinosaur, Colorado. During summer, parts of the monument can be reached over primitive roads from Elk Springs, Colorado, on U.S. 40. A few other roads penetrate the boundary of the monument.

MAPS

The Index to Topographic Maps of Colorado, obtainable free from the U. S. Geological Survey, Federal Center, Denver, Colorado 80225, shows which parts of this area are covered by topographic maps.

Hahns Peak

The general vicinity of Hahns Peak has long been known as a source of numerous quartz crystals, which occur right on top of

30

the peak itself, as well as throughout the surrounding area. The finds, however, are sporadic below the top half of the mountain. This locality was first described to the author by John C. Sampson, Jr.

The Hahns Peak, or Columbine, district, in Routt County, was once a source of gold, silver, copper, and lead from veins and bedrock, as well as gold from placers. Remains of an old flume can be seen on the south side of the peak, and in it have been found some of the best specimens. The crystals, up to 6 inches in length, occur singly and in groups in cavities up to 3 feet across. From these cavities have come the separate crystals usually picked up on the surface. They are the clear, rock-crystal variety. Limonite coats the bases of most of them, and many have patches of limonite.

This locality is accessible from Steamboat Springs by a good, gravel road (about 30 miles). You can drive to the base of Hahns Peak from the villages of Hahns Peak or Columbine; the summit (10,824 feet) is reached by a 1½-mile trail from the Royal Flush mine, on the Columbine side.

The general area was described by H. S. Gale (Ref. 1) and R. D. George and R. D. Crawford (Ref. 2).

REFERENCES

1. U. S. Geological Survey Bulletin 285, 1906, p. 28-34.
2. Colorado Geological Survey First Report, 1908, p. 189-229.

MAPS

Topographic maps: Hahns Peak (1962).
National forests: Routt.

　　　Following the broad valley of the Colorado River, Int. 70
enters Colorado from Price, Utah. Opalized wood is widely scattered on most of the low hills on both sides of the river, about
halfway across Mesa County from the Utah border. The material
found in this area varies considerably in size, quality, and appearance. The Opal Hill locality described below is perhaps typical of
the general region. Colorado National Monument and the nearby
collecting localities of Glade Park and Pinon Mesa (described
below) lie south of the river. Black petrified wood, as described
below, occurs in the Indian Hunting Ground, north of the river,
between Whitewater, in Mesa County, and Delta, in Delta County.
The highway becomes U. S. 50 southeast of Grand Junction,
where the country is characterized by barren hills and extensive
plateaus, becoming even more barren to the east. The San Juan
Mountains, described below, become visible in the distance to the

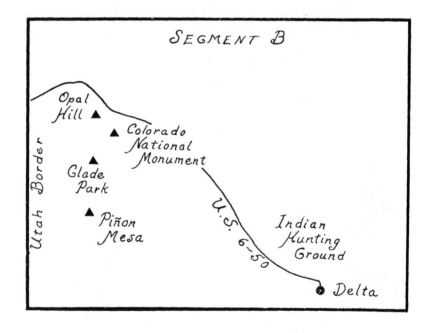

right, and the West Elk Mountains to the left. The highway goes around the southwest side of Grand Mesa, which is seen on the left most of the way. The uranium lands beckon to the southwest.

In addition to the minerals named in this portion of the book, white and colorless crystals of barite, associated with fan-shaped groups of flattened calcite crystals (including nailhead spar) occur near Grand Junction in isolated, nodular masses in sandstone, some larger than a man. Mrs. Hazel Fischer, of Grand Junction, reports these as available on open land by going about 5 miles west of Grand Junction on Int. 70 and then north toward the Book Cliffs, which form the background.

Agatized dinosaur bone, most of it fractured, was until recently rather abundantly exposed in parts of this region, but abandoned uranium mines now mark the former sites.

The States Mine Museum, still marked on some road maps, used to exhibit dinosaur tracks in coal, but is no longer in operation. Tracks of this kind are found in various nearby mines about 3.5 miles west and north of Cedaredge (follow signs to Green Valley Coal Mine); dinosaur tracks in the Red Mountain and Charles G. Gates mines are described in *Exploring Rocks, Minerals, Fossils in Colorado*, by Richard M. Pearl (Sage Books, The Swallow Press, Inc., Chicago, revised edition, 1969).

Alabaster is reported to occur 15 miles east of Delta, in pieces suitable for small ornaments but too fragile for large objects.

Collectors visiting this section of Colorado are invited to attend the meetings of the Grand Junction Gem and Mineral Club, Inc., held in the REA Building, 2727 Grand Avenue, on the second and fourth Thursdays of each month.

Kelly's Rock Shop, on the highway at Mack (Zip address 81525), specializes in conducted tours to dinosaur-bone localities and will also give free advice to collectors who want to travel on their own into rough country.

At Montrose, 21 miles southeast of Delta, the Montrose Rock and Mineral Club meets on the first Thursday of each month at the Montrose County Courthouse.

Opal Hill

Familiar to collectors in western Colorado, the Opal Hill locality was made known to others by Will C. Minor, of Fruita (Ref. 1). It has yielded at least a few hundred pounds of opalized

33

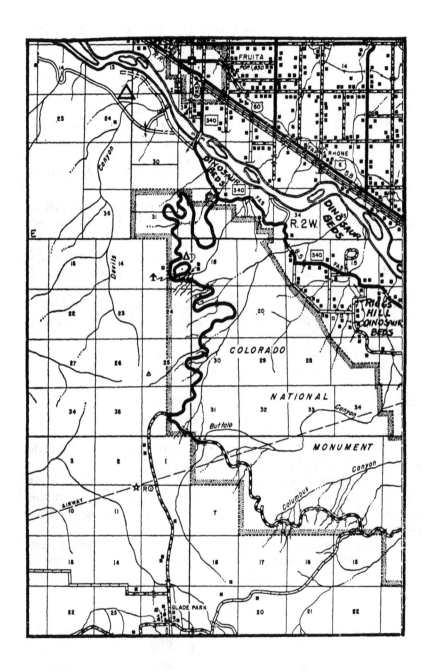

wood. Mr. Minor has found good specimens weighing 20 pounds or more. The quality and size of the material that remains at Opal Hill should not be overemphasized, but the collector is encouraged to look elsewhere in the region.

Opal Hill, better known to some residents as Blue Hill, is south of the Colorado River, 3 miles by road from Fruita. The locality can be reached from the highway intersection at Fruita according to the following log:

0.0 Junction of Int. 70 and Colo. 340. Go south on Colo. 340 toward Colorado National Monument.

1.1 Cross bridge over Colorado River.

1.4 Take gravel road to right.

1.5 Cross cattle guard onto Beard range (not marked) and cross usually dry Kodels Creek (not marked).

1.7 Continue on main road at next two forks.

2.3 Keep left at fork. Opal Hill is on the right, beyond the abandoned Fruita Golf Course, marked by the remains of a cement building.

2.9 Gate near south end of Opal Hill.

Opal Hill is a ridge several hundred feet high and about ½ mile long, trending north. Having its highest point at the southern end, it slopes generally northward toward the Colorado River, which flows past it about ½ mile from its northern end. Visible from the hill are the Book, or Roan, Cliffs to the left and Grand Mesa to the right. The town of Fruita lies just north and slightly east. The lower two-thirds of the ridge consists of pinkish shale and sandstone of the Morrison formation, of Jurassic age, overlaid by greenish shale and capped by a light-pink sandstone layer.

The likeliest place to search is in the break in the ridge, on both the east and west sides, about halfway between the highest (south) point and the Colorado River.

The opalized wood is of gem quality in the sense that it can be cut, but it does not have the play of colors characteristic of precious opal. When freshly removed from the earth, it is translucent and almost colorless, like ordinary chalcedony. Upon exposure – most of the opal has been found this way – it becomes milky white, sometimes tinged with yellow but usually plain white, having a sheen like that of opal glassware. Additional exposure causes the material first to become chalky and then to decompose entirely. Other colors, including blue, brown, yellow and green, are sometimes found. A good, green color is rare, and these specimens are chalky. Some colors occur in bands. Black, dendritic patterns of the moss-agate type are also found.

REFERENCES

1. Rocks and Minerals, vol. 14, 1939, p. 384-385.

MAPS

Topographic maps: Mack (1962).
National forests: Grand Mesa. See also the Glade Park Area map of the U. S. Bureau of Land Management.

Colorado National Monument

Although collecting is not permitted within Colorado National Monument, many specimens occur here and in the surrounding country as well. Rim Rock Drive, a 22-mile road (toll 50 cents) connects the area with both Fruita and Grand Junction throughout the year. In these spectacular 17,311 acres of beauty are colored cliffs sculptured by wind and water into peculiar shapes, which stand in striking array. Opalized and silicified wood, rock crystal, milky quartz, jasper of several colors, moss agate, banded agate, thunder eggs, amethyst, banded aragonite, rare clays, and petrified (including agatized) dinosaur bones are found in the rocks of this region, as described by W. C. Minor (Ref. 1).

From the Morrison formation, of Jurassic age, in the vicinity of Colorado National Monument have been taken skeletons of some of the largest known dinosaurs; the five species are *Antrodemus* (*Allosaurus*), *Brachiosaurus*, *Apatosaurus* (*Brontosaurus*), *Diplodocus*, and *Stegosaurus*. The best places for dinosaur bone and associated gizzard stones have been west and north of the monument, as shown on the sketch map, where faulting has lowered the productive strata several hundred feet below their altitude in the escarpment. These dinosaur discoveries are described in *Exploring Rocks, Minerals, Fossils in Colorado*, by Richard M. Pearl (Sage Books, The Swallow Press, Inc., Chicago, revised edition, 1969).

REFERENCES

1. Rocks and Minerals, vol. 18, 1943, p. 295-297.

MAPS

Topographic maps: Colorado National Monument (1962, contour or shaded-relief editions, 50 cents).
National forests: Grand Mesa. See also the Glade Park Area map of the U. S. Bureau of Land Management.

Glade Park

Opalized wood comes from Glade Park, which is reached from either Fruita or Grand Junction. The Glade Park Post Office is 18 miles from Fruita on the Rim Rock Drive, and 13 miles from Grand Junction on the same road, which now goes up No Thoroughfare Canyon. A scenic loop road 10.6 miles long connects Glade Park with two junctions on the Rim Rock Drive. This locality is on the Glade Park Area of the U. S. Bureau of Land Management.

MAPS

Topographic maps: Battleship Rock (1962), Colorado National Monument (1962, contour or shaded-relief editions, 50 cents).

Pinon Mesa

Half-a-dozen or more gem and mineral localities on Pinon Mesa have been described by Will C. Minor, of Fruita (Ref. 1). Most of them are on private ranches, on which permission to collect must first be secured, but several are on government land. Chalcedony, jasper, agate, and petrified wood of various interesting kinds are the gem materials found here.

Pinon Mesa extends from Mesa County into Utah, a distance of about 40 miles, and has a width of 10 miles. It lies south of the Colorado River and is reached by going 3.5 miles south from the post office at Glade Park, described above.

On the lower half of Pinon Mesa are found smoky-colored, white, and mottled chalcedony, and superb banded and moss agate of many colors. Mr. Minor recommends following the water pipeline up the hill and watching for specimens recently uncovered by rain. (The author did not find this pipe in 1971.) This locality is on public land in a 12-section, detached area of Grand Mesa National Forest called the Fruita Division.

Windy Point is situated at a road fork 8.3 miles from Glade Park on the main road across Pinon Mesa, past Mud Springs Camp Ground.

The left-hand road (called JS Road) leads successively to North East Creek and Johnson Creek, on the way to the south rim of the mesa. At both creeks, especially Johnson Creek, Mr. Minor reports brown and black, sugary petrified wood, some with quartz crystals in the center. Here also are black and white, banded jasper and smoky chalcedony.

The right-hand road (called 16.55 Road) from Windy Point leads, in 2 miles, to a sagebrush area on public land in Grand Mesa National Forest, near the Fruita Guard Station, where "desert roses" consisting of quartz pseudomorphs after barite have been picked up.

REFERENCES

1. Rocks and Minerals, vol. 20, 1945, p. 519-522.

MAPS

National forests: Uncompahgre. See also the Glade Park Area map of the U. S. Bureau of Land Management.

Indian Hunting Ground

Gem-quality, jet-black petrified wood has been found since the winter of 1950-51 by Will C. Minor, of Fruita, in the high-desert region between Whitewater and Delta, known as the Indian Hunting Ground. Described by Mr. Minor (Ref. 1) and examined by the author of this book, the wood takes a splendid polish; one lapidary called the solid, black material the best of that color he had ever worked with.

The outside of the wood from this locality is soft and dull, being gray or dirty tan, resembling the oil shale of the region to the north. Some pieces have the same, unattractive appearance all the way through, but most specimens are fine grained and a rich black beneath the paper-thin shell.

Although it is silicified wood, this material is somewhat softer than typical agatized and opalized wood, which is fairly common in the region. The black color is presumably due to carbon.

REFERENCES

1. Rocks and Minerals, vol. 26, 1951, p. 383.

MAPS

National forests: Grand Mesa.

URANIUM COUNTRY

Once the nation's chief source of radioactive minerals, Colorado still supplies much of the uranium of the United States. The boom of the 1950's has, however, stabilized itself at a work-a-day level that is apt to increase after the present-day contracts termi-

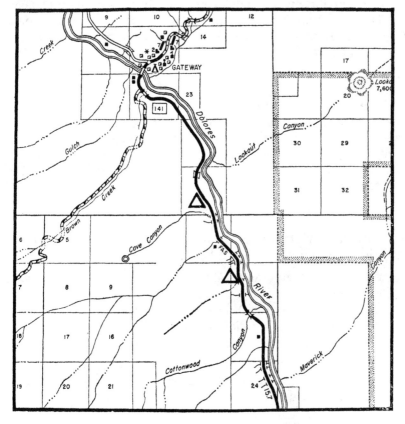

nate in the early 1970's.

The southwestern part of the state is the uranium country referred to, even though important deposits are known outside the Colorado Plateau. The unique uranium-vanadium deposits in the Four Corners region of Colorado, Utah, Arizona, and New Mexico are found mostly in the Morrison formation, of Jurassic age, and in the Shinarump and Moss Back members of the Chinle formation, of Triassic age. Those within the state of Colorado are predominantly in the upper sandstone of the Salt Wash member of the Morrison formation, of Jurassic age. Carnotite, which is bright yellow when rich, is the most familiar uranium mineral, being disseminated through sandstone and replacing or filling petrified wood in association with dinosaur bones and vegetable matter. Tyuyamunite and metatyuyamunite, coffinite, and pitchblende are other important minerals of uranium in the Colorado Plateau.

Two localities for barite and alabaster near Gateway, in Mesa County, and the occurrence of black agate over a wide area are described below.

MAPS

The Index to Topographic Maps of Colorado, obtainable free from the U. S. Geological Survey, Federal Center, Denver, Colorado 80225, shows which parts of the uranium country are covered by topographic maps. Certain Geologic Quandrangle maps are also available. The national-forest maps are Grand Mesa and Uncompahgre.

Gateway

Attractive white barite and banded alabaster occur near Gateway, in Mesa County, as logged below:

0.0 Mobil station and cafe at Gateway. Go south toward Uravan.

0.6 Cross bridge over Dolores River.

3.9 Log building in field on left across river. On right of road is old barite mine in cliff nearest road.

A vertical vein of barite about 3 feet wide was mined here. The vein continues for some distance to the right, and loose specimens are strewn around. The barite is tabular, both white and iron

40

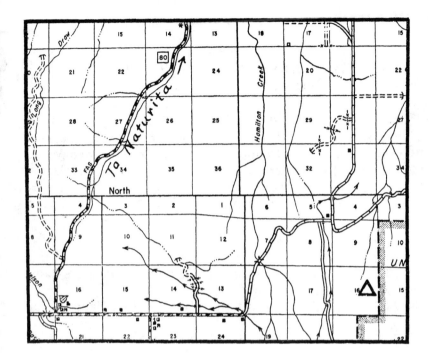

stained. It occurs in red sandstone and conglomerate of the Cutler formation, as reported by the U. S. Geological Survey (Ref. 1).

Continuing log:

6.1 Turn right to alabaster workings.

Sugary gypsum, some white and some with attractive, brown banding, occurs here in the Moenkopi formation. The alabaster in this deposit was fabricated into building blocks and other forms, near Grand Junction, as reported by George O. Argall (Ref. 2). The bed continues intermittently for ¾ mile to the right.

REFERENCES

1. *Geological Road Logs of Colorado*, Rocky Mountain Association of Geologists, 1960, p. 34.
2. Colorado School of Mines Quarterly, vol. 44, no. 2, 1949, p. 231.
3. U. S. Geological Survey Geologic Quandrangle Map GQ 55, 1955.

Topographic maps: Gateway (two scales, 1949-1960).
National forests: Uncompahgre.

Black-Agate Localities

The occurrence of black agate in western Colorado, between Paradox (in Montrose County) and Lone Cone (in San Miguel County) has been described to the author by T. Haegler, of Cedaredge. This gem material ranges up to fist size and is scattered widely on the surface. One particular locality is reached from Naturita by the following log:

0.0 Go east on Colo. 145.
4.0 Turn right on Colo. 80.
17.1 Turn left opposite Colo. 80 at store known as the Basin.
19.1 Pass school on right.
24.2 Turn right on road toward Uncompahgre National Forest.
26.2 Stop near old barn. Search east of the road across a small stream and on top of hill.

MAPS

The topographic maps are those of the southern part of the uranium country (see page 40). The national-forest maps are San Juan and Uncompahgre.

SAN JUAN REGION

The "silvery San Juans" are a magnificent group of volcanic mountain ranges in southwestern Colorado, presenting a domal structure unlike that of most of the other ranges in the state. The San Juan region has no precise limits, and each author has his own grouping and classification. The ruggedness of this region, with its towering summits and beautiful scenery, has given it the description "Switzerland of America." The San Juan region is generally considered to embrace about 12,000 square miles, but the principal mineral production has come from the 250-square-mile section lying between Ouray on the north, Silverton on the south, Ophir on the west, and the Hinsdale-San Juan county line on the east.

The mountainous nature of the San Juans is shown by the statement that San Juan County does not have a single acre of tillable soil. These mountains are highly mineralized. The mines and dumps yield specimens of almost all the important ore minerals of gold, silver, lead, copper, zinc, tungsten, manganese, and many other metals, as well as the expected gangue minerals.

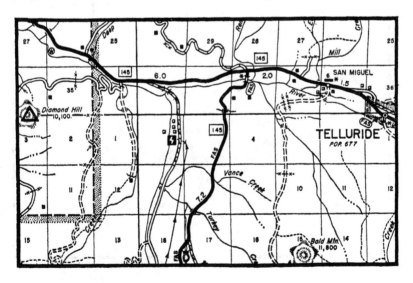

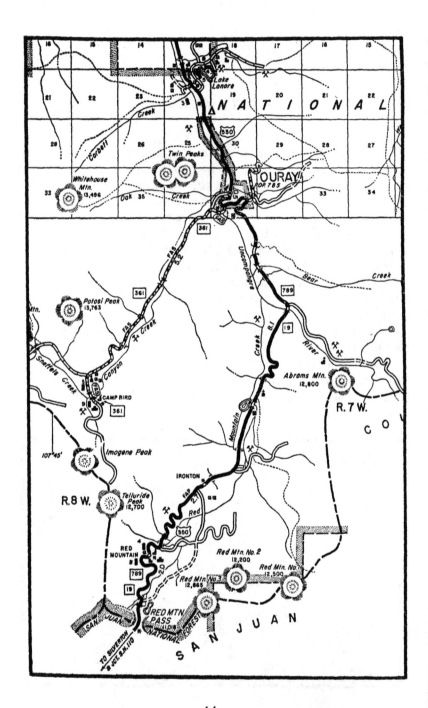

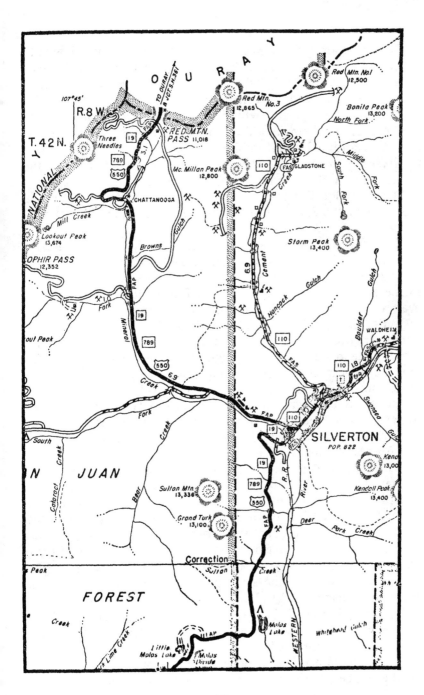

Hundreds of articles and books have been published on the mineral occurrences associated with such San Juan mining towns as Ouray, Silverton, Telluride, Rico, Ophir, Lake City, Bonanza, and Creede. The most recent comprehensive publication is by Esper S. Larsen, Jr., and Whitman Cross (Ref. 1). Summaries of the geology and mineralogy of these mining districts and references to additional literature are given by Dr. John W. Vanderwilt in *Mineral Resources of Colorado*. A useful condensation of the applicable part of that book is given in Ref. 2.

The best locality that is at present accessible to collectors in the heart of the San Juan region is the Sunnyside mill, described below.

Karl Hudson, of Durango, described many of the collecting opportunities in this most enchanting part of Colorado (Ref. 3-4). He mentioned the following specimens: crystalline masses of galena and some sphalerite near the foot of Red Mountain Pass, near Animas Forks, up Cunningham Gulch, and elsewhere; crystals of pyrite in a small gulch just south of Red Mountain Pass, on dumps southwest of Silverton, and elsewhere; chalcopyrite and tetrahedrite on a small dump about 3 miles west of Silverton; huebnerite near Chattanooga, along Cement Creek, on dumps near Silverton, and elsewhere; pyrolusite on Molas Divide about 3 miles south of Silverton; crystals of azurite and malachite at the Senorita and other small mines about 4 miles northeast of Ouray. Accessible only by Jeep is the jasper and green obsidian on Engineer Mountain.

H. A. Aurand (Ref. 5) reported the occurrence of gem chalcedony in Ouray County, but no further details are available except the statement by Frank C. Shrader, Ralph W. Stone, and Samuel Sanford (Ref. 6) that chalcedony is found in the surface gravels of that county. Mrs. Pearl found, in 1964, a large specimen of gem bloodstone along the Uncompahgre River, in the City of Ouray. Robert Keithley, of Colorado Springs, has reported (Ref. 7) the occurrence in the same place of gray and brown geodes containing quartz crystals, but the author has not found them.

Pat Fancher (Ref. 8) described the collecting of crystals and masses of enargite and crystals of pyrite at the Red Mountains, about 8 miles north of Silverton. He also mentioned (Ref. 9) various other mineral occurrences in the region.

Lawrence R. Grobl (Ref. 11) described the collecting of

46

huebnerite at the Ruby mine, near Middleton, and of enargite, chalcocite, tetrahedrite, and colusite at the Longfellow mine, at Red Mountain Pass.

According to Byron H. Rohde (Ref. 10), permission can sometimes be obtained to collect at the large dump of the famous Treasury tunnel of the Idarado Mining Company, situated near the highway (U.S. 550), 11.5 miles south of Ouray. Crystals of white "sugar quartz" occur singly (up to 8 inches long) and in huge groups, together with clusters of transparent quartz crystals. Snow-white crystals of calcite are found in association, either separately or growing upon the quartz; some fluoresce red. Fluorescent fluorite is also found here. Amethystine quartz, black-coated calcite of interesting habit, and encrustation pseudomorphs of calcite after fluorite are other popular material of late discovery. These specimens are handsome by every standard and are much in demand; some of the miners set them aside for sale to local dealers in Ouray, Silverton, and Telluride, as well as to visitors. The quartz and calcite are typical of the pockets in the veins that cut into the younger rocks throughout the San Juan region. Of the sulfide minerals, enargite crystals have been regarded as perhaps the nicest, at least until the uncovering of bright pyrite or marcasite coating fluorescent calcite (sometimes as a detachable cast), first revealed in 1969.

Benjy Kuehling, who operates the Columbine Mineral Shop in Ouray, obtained specimens of huebnerite at the Adams mine, at Silverton, in 1968-69, and also from the dump. Chris W. Christensen, of Colorado Springs, reported that fine red crystals were coming from there in 1971.

Small, terminated quartz crystals, about ½ inch in length, attractive though with brown spots, are found in virtually unlimited numbers on Diamond Hill (10,100 feet high), a bare mesa 7 miles west of Telluride, but either a Jeep or a considerable climb is required to reach them from the end of the automobile road. Inquiry may be made of Homer E. Reid at the Busy Corner pharmacy in Telluride. This locality was first made known to the author by Chester R. Howard, of Denver.

Miners at the famous Camp Bird mine, near Ouray, have supplied calcite on quartz to the specimen trade in recent times.

During the past few years, pink rhodochrosite crystals from the American tunnel of the Sunnyside mine, 8 miles northeast of

Silverton, have become well known and popular. Occasionally, the crystals are a rich red like those of the Sweet Home mine, near Alma — described (when it was accessible to collectors) in previous editions of this book. The rhodochrosite is especially noted for its fluid inclusions. Miners employed in the lower levels by the Standard Metals Corporation bring most of these lovely specimens on the market, but Gardiner E. Gregory (Ref. 12) described collecting on the one large and several small dumps; permission should be asked on the premises (road log given below). Fluorescent, green fluorite and microsize fluorite and manganocalcite also occur here, as well as other minerals of lesser interest to collectors. In 1970, miners opened up a pocket filled with manganocalcite, lighter in color than the rhodochrosite and perhaps pseudomorphs after selenite gypsum. Rhodonite from the same mine is described separately below because available under different conditions.

Other gem and mineral localities that are technically within the San Juan region are described below, along the route of Segment C, from Durango to Alamosa.

REFERENCES

1. U. S. Geological Survey Professional Paper 258, 1956.
2. U. S. Bureau of Mines Information Circular 7554, 1950.
3. Rocks and Minerals, vol. 22, 1947, p. 920-922.
4. Rocks and Minerals, vol. 23, 1948, p. 704-705.
5. Colorado Geological Survey Bulletin 22, 1920, p. 46.
6. U. S. Geological Survey Bulletin 624, 1917, p. 84.
7. Pick & Pan Reporter, vol. 2, no. 5, 1962, p. 2.
8. Rocks and Minerals, vol. 33, 1958, p. 100.
9. Mineralogist, vol. 27, 1959, p. 60-62.
10. Lapidary Journal, vol. 18, 1964, p. 73.
11. Gems and Minerals, no. 311, 1963, p. 26.
12. Gems and Minerals, no. 381, 1969, p. 20-22.
13. U. S. Geological Survey Circular 236, 1953.
14-20. U. S. Geological Survey Folio 57, 1899; Folio 60, 1899; Folio 120, 1905; Folio 130, 1905; Folio 131, 1905; Folio 153, 1905; Folio 171, 1910.
21-22. U. S. Geological Survey Geologic Quandrangle Map GQ 152, 1962; GQ 291, 1964.

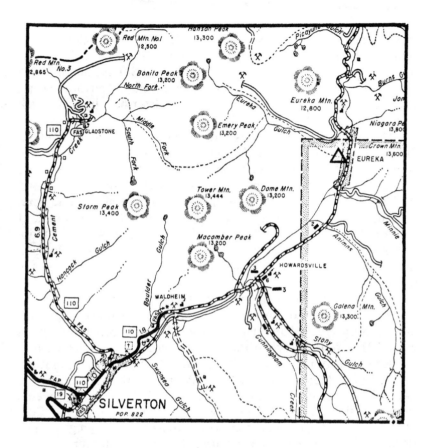

MAPS

The Index to Topographic Maps of Colorado, obtainable free from the U. S. Geological Survey, Federal Center, Denver, Colorado 80225, shows which parts of the San Juan region are covered by topographic maps. The national-forest maps of this region are Grand Mesa, Gunnison, Rio Grande, San Juan, and Uncompahgre. Certain Geologic Quandrangle maps are also available.

Sunnyside Mill

Rhodonite and rhodochrosite are the chief minerals of interest to collectors in the former Sunnyside mine, now part of the American tunnel, northeast of Silverton. The rhodochrosite crystals obtainable from miners and dealers and on the dumps are mentioned above. Because of its superior toughness — dense and

49

fine grained – a good deal of rhodonite has been cut into fine gems, as first described by H. A. Aurand (Ref. 1). This pretty, pink mineral occurs in large masses in the old Sunnyside mine and to a lesser extent in some other mines of the Silverton district, especially in the northeastern part, as described by Frederick Leslie Ransome (Ref. 2).

The Sunnyside mine was the second locality in the United States for alleghanyite and friedelite, first reported by W. S. Burbank (Ref. 3). Other minerals of interest to collectors here are helvite and tephroite, which were also described by Dr. Burbank. Alabandite and the common sulfide minerals are also found here.

The best collecting seems to be at the old Sunnyside mill, at Eureka. As mentioned by Lawrence R. Grobl (Ref. 4), gem rhodonite was the aggregate in the concrete used for the mill foundation! This is one of the most remarkable gem "deposits" in the world.

First worked for gold in 1875, the Sunnyside mine was enlarged from time to time until it became the most important mining property in San Juan County. The most productive years were 1917 to 1938, when the mine closed. It was reopened into the American tunnel in 1960 by the Standard Metals Corporation, and it has yielded much excellent material of the kind collectors like. The veins in the mine are quartz veins containing calcite, rhodochrosite, and rhodonite, as well as minor amounts of sulfides of lead, zinc, copper, and iron. These veins cut flat-lying volcanic flows of Tertiary age. The ore used to be carried over a 3-mile aerial tramway line to the mill 2,000 feet below, where the most accessible specimens are now found. The present mill is on the left side of Colo. 110, 2 miles from Silverton.

The log to this locality from Silverton is as follows:

0.0 City Hall. Go northeast past San Juan County Building, cross Animas River, bear right on Colo. 110, and continue after state highway ends. Dead-end terminus is beyond the locality. Left branch of Colo. 110 goes to American tunnel; be careful of ore trucks.

2.1 Mill of Standard Metals Corporation on left. Dump on right has ore specimens; ask permission for other collecting on nearby dumps.

4.2 Keep left at fork and go through Howardsville.

7.9 Eureka.

50

8.1 Old Sunnyside mill on left. Collecting on waste piles, especially along rail ties by cement platform.

REFERENCES

1. Colorado Geological Survey Bulletin 22, 1920, p. 58.
2. U. S. Geological Survey Bulletin 182, 1910, p. 176-178.
3. American Mineralogist, vol. 18, 1933, p. 513-527.
4. Gems and Minerals, no. 311, 1963, p. 26.
5. U. S. Geological Survey Bulletin 1114, 1961, p. 28, 35, 38, 154, 180, 283, 285-286, 325.

MAPS

Topographic maps: Howardsville (1955), Silverton (1955, small scale).
National forests: San Juan, Uncompahgre.

Going from a mountainous region to an agricultural belt, U.S. 160 passes south of the highly mineralized San Juan Mountains, previously described. From La Plata County to Rio Grande County, it goes through famed Gem Village, headquarters for mineral and gem collectors; through Pagosa Springs, with its extraordinary mineral waters; across Wolf Creek Pass, unexcelled for beauty as a place to collect specimens, as described below; and into Del Norte, which is the gateway to the important agate locality described below. A side trip on Colo. 149 takes you northwestward into the San Juans, to the historic silver-mining camp of Creede (Commodore silver mine), also described below. The former locality at Wagon Wheel Gap has been omitted from this edition because inaccessible. Instead, an amethyst locality at Beidell, north of Del Norte, has been included.

Collectors are invited to attend the meetings in Durango of the Four Corners Gem and Mineral Club, Inc., held on the first and third Monday from April to December, and the first Monday only from January to March, in its clubhouse on the alley at 24th and Main.

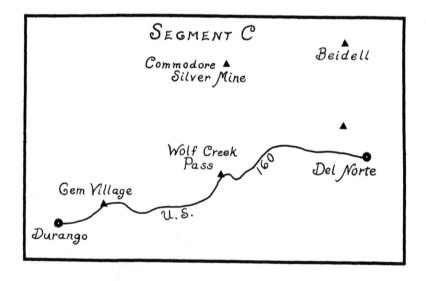

Gem Village, which is an entire community rather than just a mineral society, is given separate mention here.

Gem Village

Two miles west of Bayfield on U. S. 160 is Gem Village, a unique place founded in 1941 by the late Frank and Grace Morse. A substantial number of gem and mineral dealers, lapidaries, silversmiths, and other craftsmen and artists have settled here in the only exclusively rockhound community in America. The population is augmented each year by a cosmopolitan group of summer residents who delight in the environment. Organization is maintained through the Gem Village Museum, Inc., which elects officers.

A log book compiled by the residents of Gem Village may be consulted for information about mineral localities and tourist attractions in the region. Cafes and a motel are operated in the village, and there is a free Rockhound Campground with necessary facilities. The Navajo Trails Gem and Mineral Club, Inc., meets in its own clubhouse in Gem Village on the last Tuesday of each month, except during the winter, featuring a pot-luck supper and usually a program, to which all visitors are welcomed. A free annual show is held at Gem Village in the summer.

REFERENCES

1. Lapidary Journal, vol. 5, 1951, p. 8-10.
2. Lapidary Journal, vol. 6, 1952, p. 38-39.
3. Grit, June 22, 1952, p. 7.
4. Rockhound Buyers Guide, 1952, p. 38-40.
5. Lapidary Journal, vol. 8, 1954, p. 26-30.

MAPS

Topographic maps: Ignacio (1896-1907).
National forests: San Juan.

Wolf Creek Pass

Wolf Creek Pass crosses the Continental Divide in Mineral County on U. S. 160 at an altitude of 10,850 feet. Six miles west of the top of the pass is a well-known locality for agate nodules. Good quartz moonstone is also found at the pass and west of it, and

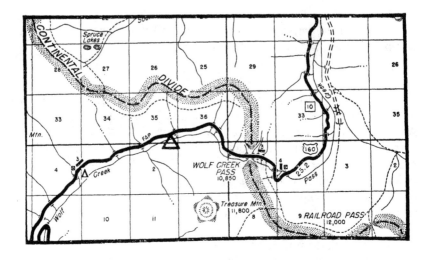

along Wolf Creek.

The agate occurs in nodules, which average up to about 4 inches across, although some of them are larger. The color is mostly white, gray, and grayish blue, and the agate is associated with white quartz crystals. Some of the crystals are outside the agate bands, and some are inside. Some nodules consist entirely of only one material or the other. The nodules, as well as the rock itself, may have a green and blue coloration, and some of the rock contains small nodules that look like turquoise but are not colored throughout. Natrolite, a zeolite mineral, occurs in nodules in the same rock. Occasional geodes contain attractive amethyst crystals.

The log to this locality from Pagosa Springs is as follows:

0.0 Junction of U. S. 160 and U. S. 84 east of Pagosa Springs. Go east on U. S. 160 toward Del Norte.

14.3 Treasure Falls. Highway begins to climb.

14.8 Stop on left side of highway. The dark cliffs on the right contain the agate in place. Other specimens are found in the steep slide rock below the road, but this is dangerous. Better specimens are said to occur in the steep rocks above the pass.

16.2 Scenic overlook ahead on left of road (described below).

The log to this same locality from the east is as follows:

0.0 Top of Wolf Creek Pass.

6.4 Scenic overlook at right of highway, giving a broad view of the valley of the West Fork of the San Juan River. Large amethyst crystals are supposed to occur somewhere up the valley seen at the extreme right. To the left is Treasure Mountain, its high cliffs said to contain hidden gold. Farther left is Treasure Falls. Still farther left is a brown cliff dipping toward the left; the agate is found here and in the loose rock below the road.

7.7 Road sign at left of highway.
7.8 Stop on right side of highway. The specimens are found in place in the dark cliffs (read description above).

REFERENCES

1. Lapidary Journal, vol. 6, 1952, p. 20-24.
2. U. S. Geological Survey Professional Paper 258, 1946.
3. Lapidary Journal, vol. 23, 1969, p. 1286-1291.

MAPS

Topographic maps: Wolf Creek Pass (1957).
National Forests: Rio Grande, San Juan.

Commodore Silver Mine

The rich silver deposits of the Creede district contain a great quantity of beautiful amethyst. Quartz is the chief gangue mineral in the mines, and includes smoky, milky, and chalcedonic varieties, as well as a good deal of amethyst, a very uncommon vein mineral elsewhere in the United States. The amethyst occurs in a variety of forms and associations, some of it being suitable for gems, although of course such material is comparatively rare. Especially good pieces have come from the famous Amethyst mine, but the dumps of the Commodore mine offer the best collecting now.

The log to this locality from Creede is as follows:

0.0 Creede Post Office. Go north on this street past Mineral County Court House and start up canyon on gravel road.
0.8 Road triangle. Right fork goes up East Willow Creek, but do not drive regular car beyond new mill; the Outlet mine, especially, has large dumps for collecting, and here too are the Holy Moses, Solomon, Ridge, Mollie S, and Monte Carlo mines mentioned below. Left fork goes to Bachelor

and Commodore mines, but do not drive regular car beyond. Driving and parking ahead are hazardous. Care should be taken on the dumps, which are very steep and dangerous to climb; this is no place for children.

The Emperius mill, situated 1 mile south of the Court House on Colo. 149 (the branch that goes to Lake City), may offer opportunities for collecting amethyst, according to Raymond F. Ziegler, of Colorado Springs.

Amethyst has also come from other mines on the Amethyst lode, which follows the Amethyst fault west of West Willow Creek.

The most attractive specimens from this locality are the banded, translucent pieces, which take a fine polish and make exquisite slabs. These are popularly known as sowbelly agate. They are even more highly prized when they contain pieces of native silver. Fortunately for present-day collectors, the amethyst was usually poor in silver content and so was discarded as waste on the dumps. This is still true, as light-weight material is being rejected for milling. The bands of amethyst usually occur next to bands of quartz of other colors, mostly blue and green hues. Some of this quartz is agate showing milky, bluish, and amethyst colors. Banded quartz is now being dyed in Japan to accentuate the colors, according to Alfred E. Birdsey, of Creede.

Amethyst crystals of a rather coarse growth are found up to several inches in length. Pale amethyst occurs in vugs in the veins, and these pockets may be removed so that such specimens somewhat resemble geodes.

During 1969 and other recent years, magnificent sphalerite, some crystallized native copper and chalcotrichite, and native silver in splendid crystal, wire, and leaf habits have been among the valued minerals of Creede production.

Sphalerite, galena, pyrite, and chalcopyrite are found implanted on intensely colored amethyst associated with green thuringite and chlorite, the color combinations being very handsome. Other minerals seen on the dumps include chlorite, barite, cerussite, malachite, anglesite, limonite, goslarite, wad, and rarely a bit of wire gold. Still other minerals were formerly mined in the upper levels. The miscellaneous occurrence of massicot at the Holy Moses mine, pyrargyrite and pyrite at the Solomon mine, sphalerite at the Ridge mine, and galena and cerussite in the Creede

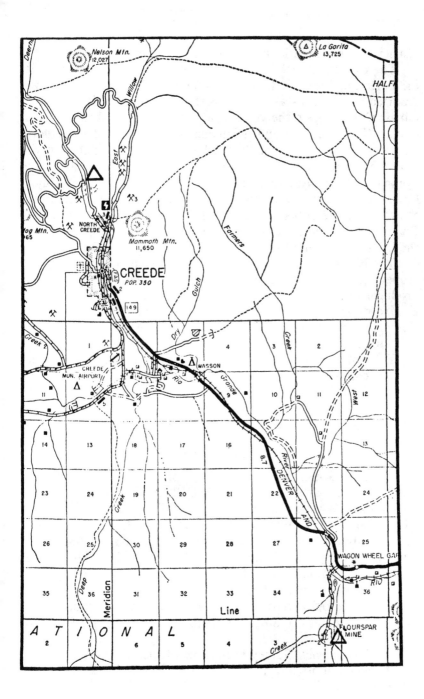

district has been mentioned by Ross Glenn (Ref. 1).

The Creede district has been described in detail by William H. Emmons and Esper S. Larsen Jr. (Ref. 2). Gem chrysoprase was reported by them as abundant in the Mollie S and Monte Carlo mines. Collecting at the Amethyst mine was described by Allan Caplan (Ref. 3). Blue and green matrix turquoise of fairly good size and quality, presumably washed down from higher in the mountains, has been picked up at several places between the Commodore and Amethyst mines, along the bed of West Willow Creek; this locality was first mentioned by Dr. Belle Katherine Stewart and described by Richard M. Pearl (Ref. 4). The Last Chance mine furnished the fine specimen of turquoise that is in the Denver Museum of Natural History.

The Emperius Mining Company took the modern leadership in reopening work at Creede, promising to make it again a major producer of silver. The potential yield has been studied by Henry C. Meeves and Richard P. Darnell (Ref. 5). The Homestake Mining Company followed with large operations, which are closed to collectors.

The splendid Creede fossils found near the town are rather similar to those of the Florissant locality, mentioned later, and are described in *Exploring Rocks, Minerals, Fossils in Colorado*, by Richard M. Pearl (Sage Books, The Swallow Press, Inc., Chicago, revised edition, 1969).

REFERENCES

1. Lapidary Journal, vol. 10, 1956, p. 53-54.
2. U. S. Geological Survey Bulletin 718, 1923.
3. Rocks and Minerals, vol. 12, 1937, p. 83.
4. *Colorado Gem Trails*, 3d ed., 1953, p. 31-32.
5. U. S. Bureau of Mines Information Circular 8370, 1968.
6. Engineering and Mining Journal Press, vol. 117, 1924, p. 973.
7. U. S. Geological Survey Bulletin 811-B, 1929.
8. Lapidary Journal, vol. 6, 1952, p. 20-24.
9. U. S. Geological Survey Professional Paper 258, 1956.
10. U. S. Geological Survey Bulletin 1114, 1961, p. 279, 302, 342.
11. U. S. Geological Survey Professional Paper 487, 1965.
12. Lapidary Journal, vol. 23, 1969, p. 1286-1291.
13. Gems and Minerals, no. 396, September 1970, p. 24-25.

Topographic maps: Creede (two scales; 1914, 1959), Creede and Vicinity (1910, 50 cents).
National forests: Gunnison, Rio Grande.

Del Norte

Plume agate is one of the more recent additions to the list of Colorado gems. The quality of the material is excellent, showing rich and varied colors and fine patterns. The choicest of this plume agate seems to be found in what is known as the Old Woman Creek area, in Saguache County, northwest of Del Norte (the county seat of Rio Grande County), at the very western edge of the San Luis Valley. There seems to be no objection to collecting at this locality, provided activity is restricted to specimens lying on the surface, but digging is prohibited. One of the claims is owned by Mrs. Juanita Davis, operator of the Del Norte Motel recommended by the Field Trip Committee of the Colorado Springs Mineralogical Society.

The log from Del Norte is as follows:

0.0 Intersection of Oak Street and Grande Avenue. Go north on Colo. 112 toward Center.
0.3 Cross Denver and Rio Grande Western Railroad.
0.5 Cross bridge over Rio Grande and turn left toward airport at junction.
1.0 Keep straight ahead at forks.
1.3 Cross bridge and turn left at forks.
2.5 Keep left at fork and pass Indian Head on right.
4.5 Branch right onto gravel road.
6.6 Turn left across gulley.
8.8 Indistinct road fork. Go as much closer either way as desired and park car near foot of Twin Mountains.

Near the parking place are ruins of a stone building once used during an earlier metal-mining operation on the hillside above it. From the slopes above this parking ground, the Sangre de Cristo Range is visible to the east, and the San Luis Valley to the southeast and south. Owing to the many mosquitoes here, it is advisable to camp overnight farther back down the road.

A profusion of agate can be found in the porous, brown, volcanic rock all over the slopes from the parking place to the base

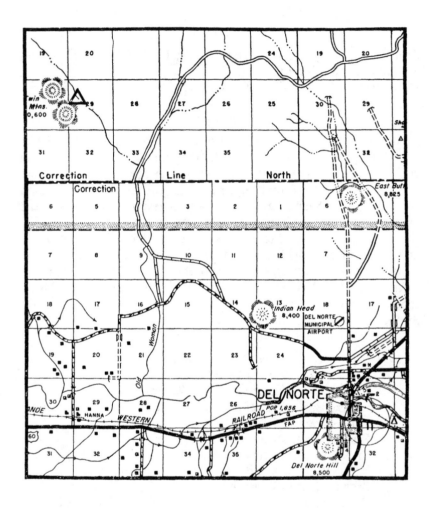

of the cliffs in the background, extending from the west to the northwest. A number of claims have been posted, and some of the productive area has been worked with power equipment. According to Rudolph Fahl (Ref. 1), this locality was first discovered by Ralph A. Dabney and Harry Simcox in 1944.

Although the plume agate is the most valuable gem occurring here, the other varieties of agate are much more common, according to David M. Seaman (Ref. 2); the dendritic and the white and colored banded agate are very attractive. Red, white, and blue chalcedony is found in solid pieces and in boytryoidal forms.

White opal, quartz crystals, and jasperized lava make worthwhile specimens. Chrysoprase, bloodstone, petrified wood, and nodules containing green inclusions are also found in this vicinity, as described by Ralph Dabney and P. L. Wilbur (Ref. 3). Besides large masses situated in place, many nodules locally called Monte Vista eggs measure up to 1 foot or more in diameter, and one has been reported to weigh 400 pounds. These include geodes, thunder eggs, and other nodules, many hundreds of which formerly lay broken on the ground.

Pieces of agate can be picked up along the road to this locality.

Farther west, on the south slopes of Twin Mountains, is found white opal having black dendritic inclusions; the locality is on private ranch property.

REFERENCES

1. Mineralogist, vol. 16, 1948, p. 34, 36-37.
2. Rocks and Minerals, vol. 27, 1952, p. 235-236.
3. Lapidary Journal, vol. 6, 1952, p. 20-24.
4. U. S. Geological Survey Professional Paper 258, 1956.
5. U. S. Geological Survey Bulletin 1114, 1961, p. 27, 276.
6. Lapidary Journal, vol. 23, 1969, p. 1286-1291.

MAPS

Topographic maps: Twin Mountains (1967).
National forest: Rio Grande.

Beidell

Crystals of amethyst quartz, typically as much as 2 inches long, come from an old gold mine at Beidell, a ghost settlement in the eastern foothills of the La Garita Mountains, about 9 miles northwest of La Garita, a largely abandoned settlement at the western edge of the San Luis Valley. A very large number of small quartz crystals of the rock-crystal variety occur here, many in clusters in cavities in the rock, from which they fall when the rock is broken. A feature of the amethyst crystals is their frequent water bubbles. Fluorescent aragonite also is found nearby in cream-colored, rounded masses. This has been a rather popular locality for mineral clubs in Colorado, in spite of the dirty nature of the deposit, the result of black manganese oxides.

The log from La Garita is as follows:

0.0 La Garita Post Office, Sinclair station, and store. Go east.
0.3 Turn left onto gravel road and continue north.
2.1 Keep left on poorer road toward Beidell Gulch.
2.5 Turn left.
5.2 Ford Beidell Gulch, the first of three times.
7.1 Cross cattle guard. Loading ramp on hill on right.
8.8 Fork right uphill on spiral road, rough but passable in regular car.
9.2 Mine and pits yielding specimens. Many rabbits around here but no drinking water; camp below mine, across main road.

This is doubtless the Carnero Creek locality mentioned in 1893 by J. S. Randall (Ref. 1), for his description of the crystals is similar: pale color, large prism faces and small "pyramids" (one is often much larger than the rest), moving bubbles. Beidell is the type locality for beidellite, a clay mineral but not one of specimen interest.

REFERENCES

1. *Minerals of Colorado*, 1893, p. 80.

MAPS

Topographic maps: Lime Creek (1967).
National forests: Gunnison, Rio Grande.

Following fast-moving streams on both sides of the Continental Divide at Monarch Pass (11,312 feet), the scenic stretch of U. S. 50 between Gunnison and Salida cuts the corners of Gunnison, Sawatch, and Chaffee Counties and passes an interesting area of pegmatite deposits situated near the noted Brown Derby mine. This outstanding property is now closed to collectors. These localities, described below, are reached from Colo. 162, which begins at Parlin, 12 miles east of Gunnison. The Quartz Creek pegmatite district, embracing an area of about 29 square miles, contains 1,803 pegmatites, which have yielded at least 27 different minerals. All except the Brown Derby are open to collectors.

North Italian Mountain, an important source of metamorphic minerals, and especially of lapis lazuli, lies north of Gunnison but is not readily accessible to the collector.

The H. W. Endner collection of minerals is exhibited in the Pioneer Museum at the eastern entrance to Gunnison. It is open daily except Sunday from 10:00 to 12:00 and 1:00 to 5:00. The late Mr. Endner was one of the early collectors in this area.

A small but bright collection of minerals and other geologic specimens is on display on the second floor of the Hurst Building at the eastern end of the Western State College campus, in Gunnison. The Clarence T. Hurst collection of Indian things, containing

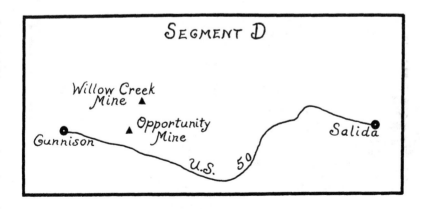

some stone artifacts, is shown upon request to the secretary, whose office is nearby.

The limestone quarry on the east side of Monarch Pass is the largest quarry in Colorado.

Advertising itself as "the Heart of the Rockies," Salida is certainly the heart of the mineral country. Few other areas of the same size in the state have produced so wide a variety of minerals, rocks, and ores as have come from within a radius of a dozen miles of Salida. The commercial products include gold, silver, copper, iron, manganese, tungsten, beryl, corundum, graphite, fluorspar, feldspar, limestone, marble, travertine, granite, and radioactive and rare-earth minerals. A list of specimens not having commercial value would be much longer.

Collectors are invited to attend the meetings of the Columbine Gem and Mineral Society, Inc., held on the second Tuesday of the month at the River Orchard Lodge in Poncha Springs; the time varies with the season.

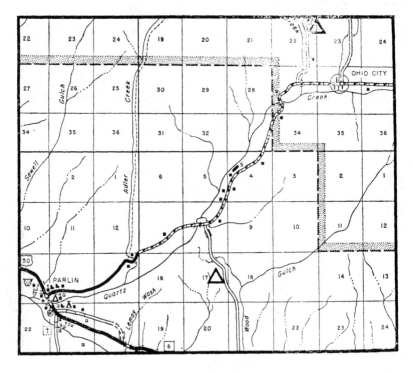

Opportunity Mine

Opened in 1939 as the Opportunity No. 1 Prospect, this mine is a group of opencut workings near Ohio City, in the Quartz Creek district of pegmatite deposits. The minerals of chief interest to collectors are lepidolite (the lithium-bearing pink mica), tourmaline, topaz, beryl, microlite (a tantalum mineral), and columbite-tantalite, in about that order of abundance.

The log from Parlin is as follows:

0.0 Junction of U. S. 50 and Colo. 162. Go north on Colo. 162, marked Ohio City, Pitkin. If coming from the east take second road in vicinity of Parlin. The road follows Quartz Creek through a fertile, flat valley much of the way, passing several cattle ranches.

4.1 Turn right off highway through gate at left curve.

4.4 Pass prospect pits on right. First located in 1948 by the Consolidated Feldspar Corporation, the one large and several small pits, which show as brown and white scars on the hillside several hundred yards away, are typical of numerous other, unnamed pegmatites in the region. Some have yielded specimens in the past. This one is barren except as an ordinary microline feldspar-muscovite mica pegmatite in gneissoid granite.

4.8 Poor road on right to Opportunity mine, 550 feet.

The deposit consists of a group of opencuts scattered on the eastern slope of a low ridge, which tends north at an altitude of 8,400 feet on the east side of Quartz Creek. The claims were located by Jesse E. Meyers and have been mapped and described by E. William Heinrich and Roswell Miller, III (Ref. 1), and by Mortimer H. Staatz and Albert F. Trites (Ref. 2).

No longer being operated, these excavations were made in some of the 11 pegmatites seen on the three Opportunity claims. The country rock is a fine-grained, reddish granite showing some banding and containing patches of biotite mica. In the granite are the pegmatites, mostly irregular in shape and ranging in size up to 730 feet long and 50 feet wide, although a barren pegmatite 3,000 feet long is exposed at the crest of the ridge, outside the Opportunity property.

Only two of the pegmatites were commercial. Four combinations of minerals are present in these deposits. One, made up

65

Milky quartz crystals, Ouray. *Katherine H. Jensen.*

Spectacular canyonlands of Dinosaur National Monument. *Colorado Advertising and Publicity Department.*

Independence Monument, an imposing monolith in Colorado National Monument. *Grand Junction Chamber of Commerce.*

Red Mountain Pass between Ouray and Silverton. *L. C. Huff, U. S. Geological Survey.*

Ouray nestles beneath the Amphitheatre in the high San Juans. *Sanborn.*

Colorado's most fascinating ghost town — Marble. *Colorado Advertising and Publicity Department.*

Lead King mill on the Crystal River near Marble. *Colorado Advertising and Publicity Department.*

Amazonstone crystal from Crystal Peak, Pikes Peak Region. *Ward's Natural Science Establishment.*

Handsome group of pyrite crystals from Leadville. *Ward's Natural Science Establishment.*

MT. ANTERO
MINERAL PARK

". WORLD FAMOUS LOCALITY
FOR SUPERB CRYSTALS OF
AQUAMARINE - PHENAKITE -
BERTRANDITE"

COLORADO MINERAL SOCIETY
1949

solely of quartz, has no specimen interest; neither does one consisting of quartz, microcline feldspar, and muscovite mica. A third is composed of beryl, topaz, and albite feldspar of the tabular variety known as cleavelandite, with small amounts of lepidolite, tourmaline, microlite, columbite, and gray quartz. The fourth contains lepidolite, quartz, and cleavelandite.

Pink lepidolite in flakes and solid blades frozen in the rock is more common in the third hole from the left, on the lower slope of the hill, where the largest and newest pits are located. Blue beryl is found mostly in the second hole from the left, in the same row of openings, but the valuable crystals were entirely removed, as far as known. Cream-colored topaz, usually coated with green sericite mica, is present in moderate amounts; at one time, rough crystals as large as 1 foot in length and 8 inches in width were obtained here.

Small masses of black tourmaline are seen everywhere, along with substantial amounts of muscovite mica and the other common pegmatite minerals: quartz and pink microcline feldspar. Interesting specimens of graphic granite can readily be picked up here by those who are intrigued by this curious rock, an intergrowth of feldspar and quartz. Albite is not ordinarily regarded as worthy of attention, but the white cleavelandite at this locality makes a most attractive matrix for the other minerals.

Columbite-tantalite was formerly recovered in crystals 2 inches in size, but available material now seems to be in minute grains only. Microlite crystals in rough octahedrons up to ¼ inch across were more abundant than columbite but are not now overly common.

REFERENCES

1. U. S. Geological Survey Professional Paper 227, 1950, p. 64-66.
2. U. S. Geological Survey Professional Paper 265, 1955.
3. U. S. Geological Survey Bulletin 1241-D, 1966.

MAPS

Topographic maps: Parlin (1964).
National forests: Gunnison. See also the Powderhorn Area Map of the U. S. Bureau of Land Management.

Willow Creek Mine

This is no longer a particularly productive locality, but its situation is accessible and most attractive, and some material of interest may still be found. When it was described under the name of Bucky Mine in the previous edition, crystals of white and pale-blue to olive-green beryl up to 1 foot in length constituted the most valuable mineral, although topaz, lepidolite, lithiophyllite-triphylite, gahnite, columbite-tantalite, monazite, and an unidentified radioactive mineral resembling samarskite occurred in smaller amounts.

The log from Parlin is as follows:

0.0 Junction of U. S. 50 and Colo. 162. Go north on Colo. 162, marked Ohio City, Pitkin. If coming from the east, take second road in vicinity of Parlin.
4.1 Pass road to Opportunity mine, described above.
6.3 Pass road to Brown Derby mine, not open to collectors.
8.5 Turn left onto dirt road. If a gate is next to the road, you have passed the correct entrance.
8.6 Keep right at fork and ford Willow Creek.
8.7 Go through gate and close it.
8.9 Pass old cabins on left.
9.3 Pass group of cabins. Mine is above on right.
9.7 Another mine is above on right.
9.8 Fossil Ridge Trail sign.

These properties, representing a total of eight claims, were worked prior to 1948 by Rod Fields and since then by the Beryllium Mining Co., Inc., and the Willow Creek Mining Co.

Zones of microcline and plagioclase feldspar, grayish-green muscovite mica, quartz, and graphic granite make up most of the deposit, as described and mapped by S. R. Wilson and W. A. Young (Ref. 1), and by Mortimer H. Staatz and Albert F. Trites (Ref. 2). John B. Hanley, E. William Heinrich, and Lincoln R. Page (Ref. 3) described this as the New Anniversary prospect as they observed it in 1943.

REFERENCES

1. U. S. Bureau of Mines Report of Investigations 4939, 1953.
2. U. S. Geological Survey Professional Paper 265, 1955.
3. U. S. Geological Survey Professional Paper 227, 1950, p. 80.

4. U. S. Geological Survey Bulletin 1241-D, 1966.

Topographic maps: Parlin (1964), Pitkin (1964).
National forests: Gunnison.

In view of symmetric Mount Sopris, Colo. 83 moves southeast-ward from Glenwood Springs, the home of copious, hot mineral waters in Garfield County, crosses the edge of Eagle County, and enters the heart of the mining area of Pitkin County. Aspen is one of the great silver camps of all time; it has recently been revived as a tourist resort for skiing and culture. Enormous masses of native silver and thick, matted wire silver, favorites among mineral collectors everywhere, have come from mines at Aspen, which include such famous names as Smuggler and Mollie Gibson. Diverging south from Carbondale on Colo. 133, the stone quarries of Marble, in Gunnison County, as described below, are reached.

In Rifle, west of Glenwood Springs, the Rifle Region Rock Club meets on the first and third Tuesday of the month at Rifle High School.

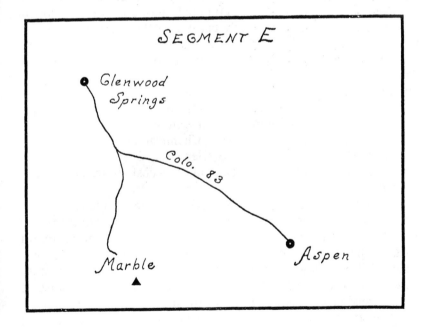

Marble

Specimens of glistening, white marble from one of the most historic and interesting quarries in the United States are available in unlimited quantities. This stone, known as Colorado Yule marble, was extracted near the community of Marble. From here came all the marble for the Lincoln Memorial, in Washington, and the first Tomb of the Unknown Soldier, in Arlington National Cemetery. For the tomb, a flawless block of 56 tons was required, and Marble was the only place in the country that could supply it. Over 60 public buildings contain Colorado Yule marble. The municipal buildings in New York City and San Francisco and the Field Building in Chicago have a good deal of the stone. In Denver, the following buildings are among many constructed of Yule marble: State Capitol Annex, Post Office, Custom House Building, City and County Building, Federal Reserve Bank, and Columbia Life Building.

Marble was recognized here as early as 1880, and the first quarry was opened in the 1890's. During the peak of 1910-1913, employment was furnished to a population of about 1,500, which dwindled to 240 in 1940, partly because of the decreased demand for marble and partly because of a fire in 1926, which destroyed the largest marble-finishing mill in the world. In 1941, enormous damage was done by a flood, resulting in the closing of the quarries, which were then owned and operated by the Vermont Marble Company. The ruins of some of the marble buildings, including the marble poolhall, can still be seen. In 1953, the property was leased by the Basic Chemical Corporation, which began in 1955 to ship stone to Glenwood Springs for use as a fertilizer and in other products, but the cost of transportation proved too high.

No log to this locality is necessary, because the many accumulations of specimens are easily seen along the highway much of the way beyond Carbondale. The gleaming piles of stone represent roadbed ballast used by the Crystal River and San Juan Railway (later Railroad) when it ran between Carbondale and Marble until 1942. Two prominent ridges of waste marble mark the course of this railroad just outside the former town.

The site of the old mill, on the edge of the Crystal River, is at the western edge of Marble. These fantastic ruins look like pictures in a book about Greece or Rome. A column of marble suspended next to the first refreshment stand is a reject from the Lincoln Memorial.

The route of the electric railroad that led from the main, or Yule, quarry a little over 2 miles away is clearly visible across the river. This quarry, with its entrance through the side of the hill, is well worth visiting for collecting and observation, but until further developing is done, it can be reached only on foot and by the local Jeep trips. Somewhat farther east is seen a parallel route to the Strauss quarry, which is on the opposite side of Yule Creek from the Yule quarry. The Strauss quarry was worked a short time in 1956.

John W. Vanderwilt (Ref. 1) described the geology of the Yule deposit and the surrounding area, known as the Snowmass Mountain area, which is in the Elk Mountains, in the extreme northern part of Gunnison County. Dr. Vanderwilt (Ref. 2) has also described the occurrence of massive beds of yellow-brown to green andradite garnet 50 feet thick, in marble, at the head of Yule Creek. Austin F. Rogers (Ref. 3) described the occurrence of braunite on Snowmass Mountain, but this is not an accessible locality.

REFERENCES

1. U. S. Geological Survey Bulletin 884, 1937.
2. U. S. Geological Survey Bulletin 884, 1937, p. 78.
3. American Mineralogist, vol. 31, 1946, p. 561-568.
4. Colorado Scientific Society Proceedings, vol. 13, 1935, p. 439-464.
5. *State Mineral Resources Board*, 1947, p. 246-247, 450-451.
6. Colorado Mining Association Mining Yearbook, 1940, p. 42-43.
7. U. S. Geological Survey Geologic Quandrangle Map GQ 512, 1966.

MAPS

Topographic maps: Marble (1960), Snowmass Mountain (1960). National forests: Gunnison, White River.

Highway U.S. 24, which joins U.S. 285 just south of Buena Vista, proceeds down the valley of the Arkansas River from Leadville to Salida, through Lake and Chaffee Counties, bordered by the highest mountains in the Rockies. Ranking sixth among all the mining districts in the United States, Leadville is an outstanding mineral producer. Gold, silver, copper, lead, and zinc are the chief metals of the Leadville ores. Pyrite for the manufacture

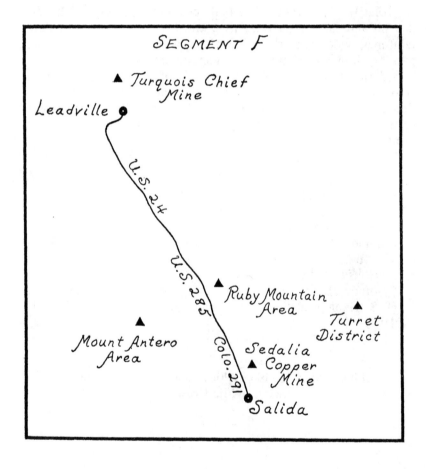

of sulfuric acid, bismuth, and iron ore and manganese-iron for metallurgical purposes are other products of this district. The innumerable mine dumps in and around Leadville furnish specimens of practically all the minerals that have been mined here; only occasional material is if any value, however, but some of this is excellent. North of Leadville are the interesting localities of Kokomo, familiar to collectors as an old source of large and bright pyrite crystals; of Climax, the site of the world's largest deposit of molybdenite; of the Alicante district, once a source of beautiful rhodochrosite; and of precariously perched Gilman, whose Eagle mine is the fourth largest zinc mine in the United States and largest copper and silver mine in Colorado. The Turquois Chief mine, described below, is west of Leadville. The Ruby Mountain area, Mount Antero area, and the Sedalia copper mine are situated along the highway between Leadville and Salida. The Turret district is reached from Salida by going back north along different roads.

Turquois Chief Mine

Worthwhile turquoise has been found at the Turquois Chief and Poor Boy Lodes, 7 miles northwest of Leadville, in the St. Kevin mining district, Lake County. Indians had apparently been aware of its existence for some time. The first systematic mining was done by two Navaho, who began working in the summer of 1935. During the two succeeding summers, they produced about 1,000 pounds of rough material, which they sent south, where it was fashioned and mounted in silver jewelry by others of the tribe.

The author had heard rumors of turquoise near Leadville, and earlier reports had placed the gem as coming from near the Mount of the Holy Cross, as reported from the Holy Cross mining district in 1890 by George F. Kunz (Ref. 1). By coincidence, a friend, Arthur J. McNair, now professor of engineering at Cornell University, and his father, Fred J. McNair, who was United States mineral surveyor for Lake County, proved to have been the original surveyors of the property for the Indians. After investigating the land and studying the geology, which was published in *Economic Geology* (Ref. 2), the author had the opportunity of acquiring the claims, but could not afford the small cost of a resurvey.

as the wartime price of turquoise soared. A bulldozer has obliter-

81

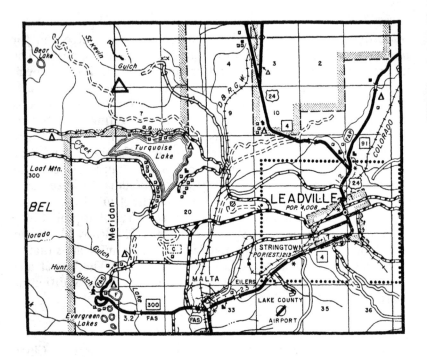

ated the former workings and drastically changed the appearance of the place as previously described in detail by the author (Ref. 3), and now little of the gem mineral seems to be left in the openings, although the dump might well be examined.

In reporting on the occurrences of uranium-bearing minerals in the St. Kevin district, C. T. Pierson and Q. D. Singewald (Ref. 4) noted disseminations of metatorbernite at the Turquois Chief mine but not in commercial amounts.

The mine can be reached from Leadville according to the following log:

0.0 Intersection of Sixth Street and Harrison Avenue. Go west on Sixth Street.
1.8 Turn right toward Turquoise Lake.
3.4 Cross Denver and Rio Grande Western Railroad.
3.5 Keep on middle branch at fork.
4.5 Fork right onto Bear Lake Truck Trail, rough road, along the high ridge that separates Turquoise Lake from drainage basin to the north.

82

to right, stopping at a cabin, from where it to walk eastward about 400 or 500 yards

is located adjacent to the Iron Mask, an older metal-producing claim. It is on the south side of a hill having a slope of 8 to 10 degrees and lying on the ridge between St. Kevin Gulch on the northwest and the gullies draining into Saw Mill Gulch on the southeast. The two highest peaks in the Rocky Mountains, Mount Elbert (14,431 feet) and Mount Massive (14,419 feet), and others in the Sawatch Range, as well as the full length of the Mosquito Range, are visible from the property and present a spectacular panorama.

The turquoise occurs in vein and nodule forms, filling faults and impregnating the wall rock, a weathered, white, medium-grained granite, which is now called the St. Kevin granite, of late Precambrian age, as named by Ogden Tweto and Robert C. Pearson (Ref. 5).

Quentin D. Singewald, who had visited the Turquois Chief mine in 1936 and 1951, has described (Ref. 6) three other turquoise deposits in this vicinity. One, the Josie May prospect, is situated 2,000 feet southwest of the Turquois Chief and has been intensively bulldozed. The others, unnamed, are situated 750 feet and 2,350 feet, respectively, northeast of the Josie May. The turquoise is associated with malachite, chrysocolla, and radioactive torbernite.

REFERENCES

1. Mineral Resources of the U. S., 1888 (1890), p. 582.
2. Economic Geology, vol. 26, 1941, p. 342-344.
3. Mineralogist, vol. 9, 1941, p. 3-4, 24-27.
4. U. S. Geological Survey Circular 321, 1954.
5. U. S. Geological Survey Professional Paper 475-D, 1964, p. 28-32.
6. U. S. Geological Survey Bulletin 1027-E, 1955, p. 289-291.
7. U. S. Geological Survey Bulletin 1114, 1961, p. 341-342.

MAPS

Topographic maps: Holy Cross (1949).

National forests: Gunnison, San Isabel, White River.

Ruby Mountain Area

Gem collectors for 80 years have found the group of three volcanic hills near Nathrop, 7 miles south of Buena Vista, to be an available source of attractive gem garnet and topaz crystals. Practically every piece of rock several inches square from favorable spots contains one or more specimens. These can be broken out with a hammer, for the rock is brittle, but it is also remarkably tough. Hence, blasting in small shots is more satisfactory and is the method used by those who have done the most successful collecting. The choicest material includes spessartite garnet of a fine, dark-red color in small but transparent and almost perfect crystals, and clear, yellow prisms of topaz.

The tiny farming settlement of Nathrop, in Chaffee County, lies on the east side of the Arkansas River valley, just west of the stream, which has been forced, by a filling of gravel, from its old course onto a granite ledge. On the opposite (west) side of the valley is the great mass of the Sawatch Range, with its glacier-scarred peaks thrust against the sky. Mount Princeton (14,177 feet high) is directly across, with 14,245-foot Mount Antero (described below) south of it. On the east are the low, rolling, forested Trout Creek Hills.

The best known of the rhyolite bodies that contain the gems, and the earliest to be studied, is Ruby Mountain, directly across the Arkansas River from Nathrop. It forms a hill about 1,350 feet long and 600 feet wide, rising about 200 feet above the river, and trending southwest. It lies just west of San Isabel National Forest. In the same direction and separated about 1,000 feet from Ruby Mountain by a dry stream bed is Sugarloaf Mountain, a larger hill about 1,600 feet long, 600 feet wide, and 375 feet high. On the western side of the river is a third mass, Dorothy Hill, about 500 feet long, 200 feet wide, and 75 feet wide.

Extensive rock quarrying was done over a period of years at Dorothy Hill. Considerable work was done more recently on the north side of Ruby Mountain since the mining for perlite for use in insulation and as a concrete aggregate has become so active in the Rocky Mountain region, but the perlite is not of commercial grade. It has been described by Alfred L. Bush (Ref. 1).

The three gem hills can be reached easily. Dorothy Hill lies

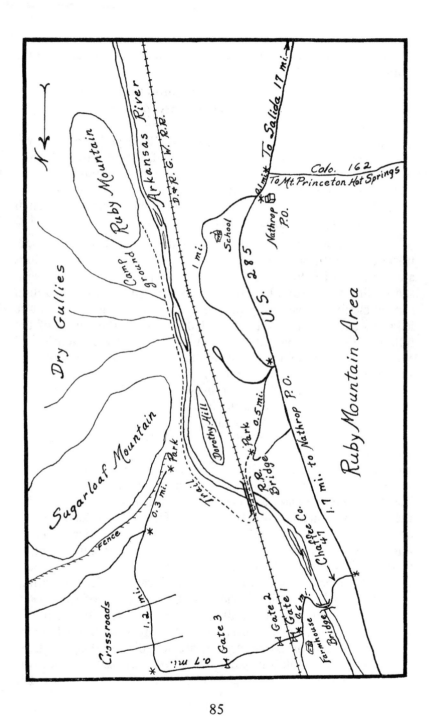

N

Ruby Mountain

Arkansas River

D. & R. G. W. R.R.

To Salida 17 mi.

Camp ground

Dry Gullies

School

Colo. 162

To Mt. Princeton Hot Springs

Nathrop P.O.

0.2 mi.

U. S. 285

1 mi.

Sugarloaf Mountain

Dorothy Hill

Ruby Mountain Area

Trail

Park

Park 0.5 mi.

0.3 mi.

R.R. Bridge

Fence

1.7 mi. to Nathrop P.O.

1.2 mi.

Chaffee Co. 47

Crossroads

Gate 3

Gate 2

Gate 1

0.6 mi.

Farmhouse

Bridge

0.7 mi.

between the highway and the river. The log to the other two hills from Nathrop is as follows:

0.0 Go north on U.S. 285 from Nathrop Post Office.
1.9 Turn right onto Chaffee County 47 (gravel). Cross the Arkansas River.
2.6 Turn right before farmhouse (on left) and cross Denver and Rio Grande Western Railroad. Trend east and then south on main road past Sugarloaf Mountain and dry gullies. Watch for posted land changes since 1971.
5.3 Campground at north end of Ruby Mountain; branch roads go in various directions.

The hills are rhyolite dikes, probably early Tertiary in age, intruded into Precambrian granite and gneiss. The bases of the hills are covered with loose rocks. Rather steep slopes alternate with sharp cliffs that drop abruptly down to the river. Ruby Mountain can be climbed most easily on the northeast side, going from one ledge to another; the uppermost ledge extends the length of the hill. Dorothy Hill has the greatest contrast between gentle and steep slopes, and the northwest side has two protuberances of rock standing above the rest of the mass.

The color of the rhyolite ranges from white to gray to pink, banded as a result of alternating lighter and darker layers. It is fine grained and contains narrow zones of quartz or enlongated cavities lined with tiny quartz crystals, which accentuate the banding. The presence of manganese seems to be the cause of the light pink or purplish tinge of the rhyolite, seen especially when it is wet.

The upper part of Ruby Mountain is composed of this rhyolite, but along the base, especially at the north and southeast sides, there are large masses of glassy perlite. The perlite is lustrous gray and contains many rounded pellets of black obsidian properly known as marekanite. These are popular with mineral collectors, who know them as "river pearls" or "Apache tears," although they are called "black rubies" by the local people, the garnets being termed "red rubies." They can be rolled out of the cavities and picked up by the thousands. They cut into attractive gems.

Around the north and east sides of Ruby Mountain, along the partly filled bed of a dry stream, are low-dipping beds of light-gray, white, and pink ash, particularly well exposed on the lower slope of the east side. Sugarloaf Mountain and Dorothy Hill contain much rhyolite with a porphyritic texture, the phenocrysts being small smoky-quartz and glassy sanidine crystals.

Countless spherical and elliptical cavities are scattered throughout the rock. They are called lithophysae ("stone bubbles") and were first described from another locality by von Richthofen in 1860. In the rhyolite, the concentric overlapping shells, ranging in diameter up to about 2 inches, are lined by drusy crystals of sanidine, with quartz upon them. The sanidine, which is a colorless, glassy, sodic variety of orthoclase feldspar, is interesting in that it exhibits a satiny luster when held in certain positions, and a blue color parallel to a particular face when magnified. The quartz is clear and colorless to gray, showing striated prisms and doubly terminated pyramids. The thin, black coating in the cavities may be an oxide of maganese.

The most common gem mineral is garnet, and several crystals may grow on different sides of the same cavity. The garnet is the manganese-aluminum subspecies called spessartite. The crystals are deep red to cinnamon red, the darker colors occurring in the more evenly grained rhyolite. They are transparent and range in size up to about ½ inch, although most of them are hardly over a quarter that size.

The crystals of topaz are usually a fine, yellow color when freshly obtained from a solid rock mass, but they seem to fade to pale yellow, bluish, or quite colorless upon exposure to light. Most of the crystals are about 3 millimeters long, although some have been found as large as 12 millimeters (about ½ inch), well suitable for cutting. The crystals are prismatic, and the tiny ones in Sugarloaf Mountain and Dorothy Hill look like quartz. They are attached to the cavities in different positions, so that double terminations are common.

REFERENCES

1. Colorado Scientific Society Proceedings, vol. 15, 1951, p. 326-327.
2. Rocks and Minerals, vol. 22, 1947, p. 109-110.
3. Mineralogist, vol. 7, 1939, p. 359-360, 388-389.
4. American Journal of Science, ser. 3, vol. 31 (131), 1886, p. 432-438.
5. Colorado Scientific Society Proceedings, vol. 2, pt. 2, 1887, p.

61-70.

6. American Mineralogist, vol. 22, no. 12, pt. 2, 1937, p. 13-14.
7. Mineral Collector, vol. 1, 1894, p. 131-133.
8. American Journal of Science, ser. 3, vol. 47 (147), 1894, p. 390, 391.
9. U. S. Geological Survey Bulletin 1114, 1961, p. 16-17, 139, 159-160, 275, 333.

MAPS

Topographic maps: Buena Vista (1955), Poncha Springs (1956).
National forests: San Isabel.

Mount Antero Area

The beryllium and other minerals that occur on Mount Antero and in the surrounding area of central Chaffee County make this one of the most interesting mineral localities in the United States, as well as the most interesting gem deposit in Colorado. It is also the highest gem locality in North America and third highest in the world.

Antero is not for the casual collector. It is a long, hard climb – not at all dangerous, but always strenuous. Local conditions are not the kind to encourage commercial mining except when more remunerative than here. The deposits occur at an altitude of about 14,000 feet and are accessible only during the short summer. Frequent, severe electrical and hail storms hamper operations, and the steep, talus-covered slopes of the mountains are conducive to landslides. Supplies and equipment are transported with difficulty. But the thrill of the scenery – unique among American mineral localities – and the beauty of the wild life and flowers make up for all the hardships.

The locality is especially worthwhile for the serious collector and prospector, because the mineral-bearing cavities are spread over a rather large area and the probability of uncovering a new deposit is not too remote. Many prospect holes can be seen in the area, some of them representing separate periods of development. Apart from the recent activity of the Antero Mining Company, most of the workings are pits several feet in diameter and about the same depth, opened by means of simple hand tools and a small amount of blasting powder. Except for an occasional summer

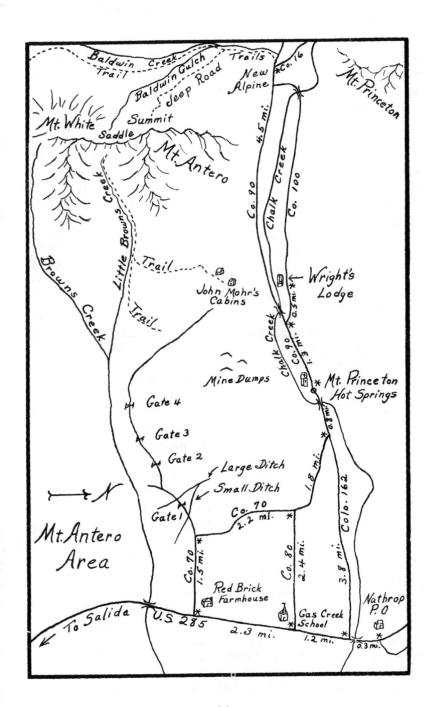

Baldwin Creek Trail

Baldwin Gulch Jeep Road

Trails

New Alpine

Co. 16

Mt. Princeton

Mt. White

Summit

Saddle

Mt. Antero

Co. 90

Chalk Creek

4.5 mi.

Co. 100

Little Browns Creek

Trail

Trail

Browns Creek

John Mohr's Cabins

Wright's Lodge

0.3 mi.

Co. 90

1.3 mi.

Chalk Creek

Mine Dumps

Mt. Princeton Hot Springs

Gate 4

0.8 mi.

Gate 3

Gate 2

Large Ditch

Small Ditch

1.8 mi.

N

Gate 1

Co. 70
2.2 mi.

Colo. 162

Mt. Antero
Area

Co. 70
1.5 mi.

Co. 80
2.4 mi.

3.8 mi.

Red Brick
Farmhouse

Nathrop
P.O.

To Salida

U.S. 285

2.3 mi.

Gas Creek
School

1.2 mi.

0.3 mi.

mining claim, the land is entirely public. The large-scale mining operations that began in 1956 have now ceased, and the road that was built practically to the summit is accessible (1971) only to the Jeep-type vehicle.

The discovery of aquamarine at Mount Antero is credited to Nathaniel D. Wanemaker in 1884 or 1885, as reported by R.T. Cross (Ref. 1). Local rumors say that others, including Tom Ivey, Jasper Pettigrew, and John Mohr, collected crystals in earlier years, as seems likely. George F. Kunz (Ref. 2) published the first notice of Wanemaker's discovery in 1887, giving the locality as the Arkansas Valley.

Systematic prospecting began in 1888, the yield for that summer being valued at $600 or $700 worth of crystals, from which gems up to 12 carats in weight were cut. By 1892, $5,000 worth of gems had been cut, according to Dr. Kunz's reports. J. D. Endicott worked the deposits actively in 1909 and 1910.

Edwin W. Over, Jr., had been the chief and most successful individual producer at Antero in recent years, maintaining a camp here during several summers, including 1928, 1931, 1932, 1933, 1938, 1951, and 1953 — a total of 12 months' working time. A crystal found by him in 1932 and now in the Harvard Mineralogical Museum is 7 inches long and 1¼ inches across, being one of three aquamarine crystals of that length recovered by him from this locality. Mr. Over described some of his activities here in several articles (Ref. 3-4). Arthur Montgomery, a companion of his on one expedition, wrote "Storm Over Antero" (Ref. 5), which conveys better than any other report the true fascination of this place.

The Colorado Mineral Society maintained a summer camp in 1938 in the canyon on the north side of Little Browns Creek at an altitude of about 12,000 feet, close to timberline, where water, fuel, and shelter are obtainable. Acting upon a suggestion made in 1941 by Chester R. Howard, at that time president, the society in 1949 sponsored the erection on Colorado Day (August 1) of a bronze plaque in recognition of the establishment of Mount Antero Mineral Park. Nine members, under the leadership of James R. Hurlbut, reached the top of the peak that day, and the plaque was mounted in a large granite pinnacle south of the summit.

The deposits associated with the name Antero occur in pegmatites and veins, mostly in a body of white Antero granite, of

Tertiary age, which underlies a large part of Mount Antero (14,269 feet high) and White Mountain (13,347 feet) and also appears about ¾ mile west of these peaks. Some of the same deposits are found in an older gray quartz monzonite, which belongs to the same Princeton batholith but which surrounds the granite; these pegmatites and veins, however, are always within a short distance of the contact, having arrived with the invading granite. In addition, there is the California vein (described below), which is in the quartz monzonite about 2 miles southwest of the top of Mount Antero, at the foot of Mount Baldwin.

Mount Antero and White Mountain are connected by a broad ridge, which forms a distinct north-south divide. Mount Baldwin (mapped as Carbonate Mountain and several times referred to as Calico Mountain by confusion with a mountain of that name situated about 3 miles southwest) stands west of the divide between Baldwin Creek and East Baldwin Creek.

These peaks belong to the Sawatch Range, which extends for about 75 miles north and south between the Eagle River on the north and Tomichi Creek on the south. Although the Continental Divide follows the Sawatch Range for most of its length, the mountains here lie somewhat east of it.

The large streams having their sources in the chief elevations of the area flow eastward in deep canyons to the Arkansas River. Chalk Creek flows between Mount Antero and Mount Princeton, north of Antero; Little Browns Creek flows between Mount Antero and White Mountain, south of Antero; and Browns Creek flows between White Mountain and Shavano Peak, still farther to the south.

Because this area is a difficult one to traverse, the accompanying map is given as an aid, but the newcomer should certainly make local inquiry, especially regarding freedom of access. The nearest settlement is Nathrop, and a number of ranches and camps are in the region. The approach by trail up Little Browns Creek from Nathrop, 7 miles south of Buena Vista, is logged below. A transverse approach is from Wright's Lodge, near Chalk Creek (where instructions and pack horses can be obtained), by way of the Baldwin Gulch trail, which begins about 1 mile west of New Alpine, on Colo. 162. The recent mining road takes off from Chaffee County 90 about 6 miles east of the ghost town of St. Elmo, branching to the left where the Baldwin Lake road goes to the right. This road also goes up White Mountain.

The log from Nathrop is as follows:

0.0 Go south on U. S. 285 from Nathrop Post Office.
3.8 Turn right onto Chaffee County 70 (gravel) beyond red brick farmhouse.
5.3 Turn left where main road turns right, and cross small irrigation ditch. The mine road (dirt) begins here.
5.5 Go through gate and cross bridge over large ditch. Close all gates.
5.9 Keep right at fork.
6.4 Go through second gate.
6.7 Go through third gate.
7.2 Go through fourth gate.
9.8 Reach cabins formerly occupied by John Mohr at base of Strawberry Mountain, at about 9,000 feet elevation. Then ride horseback or hike up the trail, which climbs over 5,000 feet in 7 or 8 miles. The camping ground previously mentioned as occupied by the Colorado Mineral Society is a satisfactory place to stop, although others equally good are farther upstream.

The topography is rugged in the extreme, with a succession of high mountains and deep canyons bordering on a broad valley. The valley of the Arkansas River lies below, stretching 23 miles north and south between Buena Vista and Salida, and east from the edge of the range to the Trout Creek Hills, on the other side.

The summit area of Mount Antero is only a few square yards, but there are great ridges extending north, east, and south. The north ridge is precipitous and drops sharply into Chalk Creek Canyon. The east ridge runs straight for about 1 mile. The south ridge form a narrow depression known as "the Saddle," which rises to a knob that entirely conceals the top of Mount Antero when it is approached from the south.

White Mountain, like Mount Antero, reaches its highest elevation at its western end; it really has two summits a few hundred yards apart and of about equal height. From them, a ridge extends eastward and branches off into two ridges, the southerly one descending more rapidly than the other. Mount Baldwin has a distinctive, symmetric profile on the east. Cirques and small lakes show former glaciation in the area.

The general geology of this region was described by R. D. Crawford (Ref. 6). George Switzer (Ref. 7) has classified the pegmatites of the Mount Antero region on the basis of their mineralogy as follows:

1. Beryl pegmatites: Beryl-smoky quartz.
2. Phenakite pegmatites: Phenakite-colorless quartz. Phenakite-smoky quartz.
3. Beryl-phenakite-bertrandite pegmatites: Beryl-phenakite-bertrandite. Beryl-phenakite-bertrandite-fluorite.
4. Topaz pegmatites.

The gemmy crystals occur in large miarolitic cavities in the upper 500 feet of the granite, on the higher slopes of White Mountain and Mount Antero (practically to 14,000 feet). The pegmatites in this productive zone are roughly disk shaped or cylindrical and are mostly smaller than 3 feet in length. Some of the cavities are close together, while others are isolated, and there is no regular arrangement. Where they are concentrated, a digging such as the Antero Lode may include more than a single pocket. Other pegmatites, which are much more widely distributed through the granite, are nearly solid, consisting almost entirely of beryl. Quartz veins containing beryllium minerals and some molybdenite, huebnerite, and sanidine feldspar, but little microcline or albite feldspar, are present in the immediate area.

The finding of good specimens seems to be more or less accidental, because the steep slopes in this region are covered with talus, and the unbroken mountain mass is exposed only in a few places. Successful discovery at one spot is naturally followed by further search at the same place, so the diggings are quite localized. The best finds of aquamarine have probably been on White Mountain, along the ridge close to the summit. The Saddle of Mount Antero has also been a good locality; here is the Antero Lode worked in 1938 by Edwin W. Over, Jr., and Arthur Montgomery. This digging marks about the northern limit of the area that has been profitably worked on Mount Antero in the past, although indications of beryl are present everywhere. The west slope of Mount Antero seems to be less productive than the other sides.

The common beryl of the Antero region has a fairly bright, blue color. Most of the crystals are badly fractured, apparently as a result of freezing and thawing of ground water, rendering them opaque. The crystals range in length up to about 8 inches, although the average one that would be called large might be from 2 to 3 inches long. They are typically long, hexagonal prisms, distinctly striated lengthwise. They are etched in a fashion that was first noticed on specimens from here and described by R. C. Hills (Ref. 8) in 1889.

Aquamarine, the gem variety, is usually blue to pale blue green at Antero but some specimens are the prized, deep-blue color. The aquamarine occurs on colorless and smoky quartz, muscovite and biotite mica, feldspar, and a clay mineral (probably kaolinite altered from the feldspar), and it is found in loose crystals from which the clay material has apparently been removed.

In addition to beryl, two other beryllium minerals — phenakite and bertrandite — are found. The Antero bertrandite is among the finest in the world; it is intimately associated with beryl and occurs in colorless, tabular crystals and heart-shaped twins, in many places implanted upon crystals of aquamarine and enclosing phenakite.

Phenakite is second only to aquamarine in interest among the minerals of the Antero pegmatites. Much of it is of gem quality, and Antero is the most important locality in the United States. The first specimens were found in 1886 by Nathaniel D. Wanemaker, who gave them to R. T. Cross, who in turn made them available to Samuel L. Penfield (Ref. 1, 9) for study.

The following summer, Wanemaker revisited the place, accompanied by Walter B. Smith; they found two localities near the summit of Mount Antero, which yielded several hundred specimens that sold for over $500. The largest phenakite crystals from Antero measure 2 inches in diameter. Most are opaque, but many have transparent spots, and some, particularly the twins, are perfectly transparent and pale yellow in color. The simple crystals are white and colorless to faintly bluish gray. Both the prismatic and rhombohedral habits are represented. The phenakite occurs more with fluorite, especially green octahedrons, and with quartz than with any other minerals. Crystals of quartz have been found with phenakite crystals enclosed in their centers. Some phenakite with bismutite implanted upon it has been described by F. A. Genth (Ref. 10). Phenakite and aquamarine are not normally found together here.

Rock crystal (clear quartz) is abundant in the pockets at

Antero. Bright crystals several inches long are also found lining openings in the large veins of milky-white quartz that are so numerous. The specimens have been famous for many years for their fine quality. Flakes of hematite coat some of the crystals as "dust." In 1892, George F. Kunz (Ref. 11) mentioned a large sphere, slightly less than 6 inches in diameter, cut from Mount Antero rock crystal and shown in the Mines and Mining Building of the World's Columbian Exposition; it was not perfect, he said, but "quite equal to the crystal balls of the 18th century." This may be the same rock-crystal sphere as the one now in the gem hall of the Field Museum of Natural History and described by Dr. Clifford C. Gregg as "one of the largest – 5½ inches in diameter."

Smoky quartz is the most common mineral in the Antero pegmatite bodies. Crystals have been found weighing nearly 50 pounds. Many of them bear crystals of aquamarine and phenakite upon them, and phenakite crystals are occasionally found, both enclosed within the smoky quartz and resting upon it. Many of the smoky-quartz crystals are, of course, opaque or fractured, with clear material only in spots. George F. Kunz mentioned in 1892 a beautiful faceted gem 84 millimeters (over 3 inches) long, found in 1891 and displayed by Tiffany and Company at the World's Columbian Exposition.

The feldspar of the cavities is mostly microcline, with some albite, and occurs massive and in simple crystals and Carlsbad and Baveno twins, some of which weigh several pounds each.

Extremely beautiful fluorite comes from the Antero pockets. It ranges in color from pale green to purple, most commonly the latter, and occurs in octahedrons, combinations of octahedrons and dodecahedrons, and rare twins. It has been found in single crystals up to 6 inches across and in much larger clusters.

Topaz has been reported almost since the earliest development of the area; it is found in irregular crystals, some attached to the walls and others lying loose in the cavities.

Muscovite and biotite mica are found in most of the pockets. Goethite pseudomorphs after pyrite are fairly well distributed through the cavities, and specimens 1 or 2 feet across have been collected. Other minerals that have been reported from the Antero cavities include sulfur (in scattered grains and tiny, indistinct crystals with limonite), spessartite garnet (discovered by the author, Ref. 12), apatite, ilmenorutile (mostly massive), monazite, bismutite (on phenakite), hematite (small crystals), molybdenite (single flakes and groups), molybdite, brannerite (a radioactive mineral), cyrtolite, pyrite, tourmaline, sericite, calcite, magnetite, perhaps columbite or tantalite, and a green mineral resembling serpentine or talc.

White Mountain mentioned above is now called Mount White.

Some aquamarine and colorless beryl have been found on the dump of the California molybdenum mine, which is situated on the north side of Browns Creek and near its head, at an altitude of about 12,500 feet. A Jeep road connects it with the Antero-White deposits. Beryl crystals up to 2 inches long occur here. Dr. Kenneth R. Landes, of the University of Michigan, reports that a number of small, clear crystals were obtained during the mining of the property by the Molybdenum Mines Company, of Denver, in 1917 and 1918. The deposit has been described in some detail by Philip G. Worcester (Ref. 13), by Dr. Landes (Ref. 14), and by John W. Adams (Ref. 15). A brief report of recent mining for beryl was given by Henry C. Meeves (Ref. 26). The vein minerals include quartz, beryl, molybdenite, ferrimolybdite, jarosite, muscovite mica, tourmaline, fluorite, rutile, magnetite, and brannerite.

REFERENCES

1. American Journal of Science, ser. 3, vol. 33 (133), 1887, p. 132-134, 161-162.
2. Mineral Resources of the U. S., 1885 (1887), p. 439.
3. Rocks and Minerals, vol. 10, 1935, p. 27-29.
4. Rocks and Minerals, vol. 3, 1928, p. 110-111.
5. Rocks and Minerals, vol. 13, 1938, p. 355-369.
6. Colorado Geological Survey Bulletin 4, 1913.
7. American Mineralogist, vol. 24, 1939, p. 791-809.
8. Colorado Scientific Society Proceedings, vol. 3, 1889, p. 191-192.
9. Colorado Scientific Society Proceedings, vol. 2, 1887, p. 144-146, 177-179.
10. American Journal of Science, ser. 3, vol. 43 (143), 1892, p. 188-189.
11. Mineral Resources of the U. S., 1892 (1894), p. 766.
12. American Mineralogist, vol. 26, 1941, p. 54.

13. Colorado Geological Survey Bulletin 14, 1918, p. 34-38.
14. Economic Geology, vol. 29, 1934, p. 697-702.
15. U. S. Geological Survey Bulletin 892-D, 1953.
16. American Journal of Science, ser. 3, vol. 40 (140), 1890, p. 488-491.
17. American Journal of Science, ser. 3, vol. 36 (136), 1888, p. 52-55, 317-331.
18. Mineral Resources of the U. S., 1888 (1890), p. 588.
19. Mineral Resources of the U. S., 1909, pt. 2, 1911, p. 748.
20. American Journal of Science, ser. 3, vol. 44 (144), 1892, p. 387.
21. Rocks and Minerals, vol. 32, 1957, p. 148-150.
22. American Journal of Science, ser. 3, vol. 37 (137), 1889, p. 215-216.
23. Chemical Review, vol. 5, 1928, p. 17-38.
24. American Mineralogist, vol. 37, 1952, p. 910-930.
25. U. S. Geological Survey Bulletin 1114, 1961, p. 16, 69-70, 71, 78, 82, 122, 150, 159, 252-253, 275, 333.
26. U. S. Bureau of Mines Report of Investigations 6828, 1966.
27. Lapidary Journal, vol. 21, 1967, p. 85-93.
28. Lapidary Journal, vol. 22, 1968, p. 1054-1058.
29. U. S. Geological Survey Professional Paper 501-A, 1964, p. A9.

MAPS

Topographic maps: Garfield (1940), Poncha Springs (1956).
National forests: Gunnison, San Isabel.

Sedalia Copper Mine

The abandoned Sedalia copper mine, 4 miles north of Salida, at an altitude of 7,900 feet, in the Trout Creek Hills, is a famed locality for huge almandite-garnet crystals of remarkable perfection. Apparently unlimited in quantity, these garnets require some exertion to collect but no special skill or equipment. A suite of them, of varying sizes but otherwise alike, makes a handsome display. More than a dozen minerals of lesser interest can also be secured at this place.

There are three main kinds of minerals, as follows:

1. Garnet and other minerals in the schist country rock.

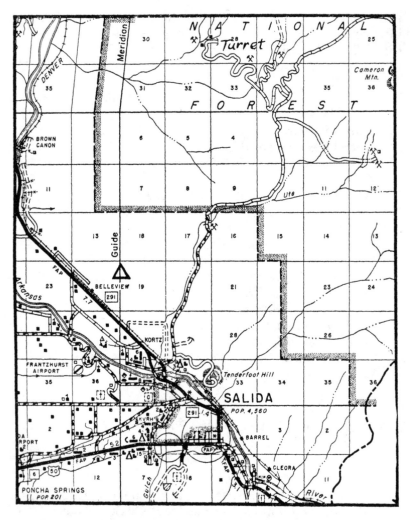

2. Minerals in the copper ore body.
3. Pegmatite minerals.

Early in the century, the Sedalia mine was one of the few important copper mines in Colorado. It was discovered in 1884 and worked intermittently for a number of years, declining after about 1910. Three tunnels were opened, and a leaching and centrating plant was operated 1 mile west of the mine, as originally described by Waldemar T. Lindgren (Ref. 1). Some silver and a trifling amount of gold were recovered from sulfide ores. After a long period of abandonment, the property was tested unsuccessfully for uranium in 1953 by the Empire Zinc Corporation of the New Jersey Zinc Company. E. E. Lewis, Inc., and International Resources, Inc., have worked here since.

The log from downtown Salida is as follows:

0.0 Intersection of Main Street and Colo. 291. Go north on Colo. 291 toward U. S. 285. Cross Denver and Rio Grande Western Railroad.

1.0 Pass Spiral Drive (Chaffee County 180) to right. (This road leads to the Turret district, described below.)

1.3 Cross bridge over Arkansas River.

4.0 Turn right, opposite Chaffee County 35 (not marked in 1971), and cross railroad tracks. Turn immediately right, following road along tracks past two farmhouses.

4.6 Turn left through gate and close it. Go toward hills and follow along fence.

5.5 Turn right through second gate and close it.

6.4 Park near mine.

The country rock is mostly an ancient schist, a metamorphic rock of Precambrian age. Successive zones of gneiss, biotite schist, staurolite schist, actinolite schist, chlorite schist, and quartz schist occur in the vicinity of the mine (Ref. 2). In the chlorite schist are found the large garnets; they are abundant enough in places to give the name garnet schist to the rock. Inasmuch as this general type of green, thin-layered rock is exposed over a considerable area of these hills, such garnets should be plentiful for countless centuries to come. They seem to be especially numerous adjacent to the main pegmatite dike, which splits the ore deposit in two, but they do not seem to have been the result of this intrusion, and so the association is evidently a coincidence.

Garnet crystals as heavy as 15 pounds have been taken from this locality. Being iron-aluminum garnet, they belong to the subspecies almandite. They are dodecahedrons, many practically perfect in shape; they are almost always altered on the outside to green chlorite (aphrosiderite), which can be scraped off to reveal a brown interior (Ref. 3). Since the chlorite penetrates to different depths, however, it is recommended that the specimens not be trimmed but left instead as found, either coated green or in their original, unaltered, brown color. The crystals are homogeneous inside but usually not very solid, crumbling if struck or often even if dropped. These excellent crystals may be secured in a complete gradation in sizes from tiny grains to individuals several inches in diameter. Placed side by side in a mineral cabinet, they present a fine appearance.

Grown in the chlorite and attached to the faces of the garnet are octahedrons of magnetite. Actinolite, asbestos, talc, and small quartz crystals can also be collected in the schist. Other typically metamorphic minerals in the schist include sillimanite, biotite, epidote, feldspar, corundum, spinel (gahnite), staurolite, and hornblende and other amphiboles (glaucophane?), but they are closely intergrown in the rock and are not of specimen quality. The occurrence here of green, radiating crystals of kyanite and green, chromium-bearing mica have been mentioned by Ross Glenn (Ref. 4).

Garnet is obtainable on the surface or in the old underground workings. The best results in the open are to be had by climbing the steep hill to the highest and largest dump. A productive area of chlorite schist, yielding garnet and magnetite, is situated about 100 feet back (east) of the north end of this dump. Unused mine tunnels anywhere are always potentially dangerous.

The former ore body is a thick bed of actinolite schist impregnated with copper minerals in a flat, lens-shaped mass. On a ridge about 600 feet above the valley, the ore is exposed as soft, whitish rock stained in spots by copper colors.

Among the copper minerals found in the various openings and on the dumps are chalcopyrite, chrysocolla, malachite, azurite, cuprite, chalcanthite, and chalcocite. The chalcopyrite is primary, the rest being the result of oxidation. Chalcocite and malachite are the most important. Copper-bearing pyrite and copper-colored chalcedony quartz (resembling chrysoprase) are also found as primary minerals. Chrysoprase of apple-green color has been reported on the dump by John W. Adams. Sphalerite and galena having the same origin are found here, especially underground. Melanterite and limonite (iron minerals), hemimorphite and willemite (zinc silicate minerals), and cerussite (lead carbonate) were produced by the alteration of the ore body and are available as specimens. The Sedalia mine is one of the few known sources of willemite, and it is rare here, being observed only in tiny crystals that are colorless and clear. When first discovered and described in 1894 by Samuel L. Penfield (Ref. 5), this willemite represented the third American locality for the mineral. The presence of pyrophyllite in huge quantities was reported by Lyle E. Nesbitt (Ref. 6). He also noted black to brown to blue rare-earth minerals on the edge of the dumps near the top.

The deposit is penetrated by a wide pegmatite dike, which yields beryl, microcline feldspar, quartz, and muscovite mica directly above the mine. Other pegmatites, of little interest, are situated both north and south of the mine. The earlier interpretation of this body as a regularly metamorphosed, basic, magmatic segregation has been doubted by Charles A. Salotti (Ref. 7), who believes it to be a sulfide deposit of the magnesium-aluminum silicate gangue type.

REFERENCES

1. U. S. Geological Survey Bulletin 340, 1908, p. 161-166.
2. Colorado Scientific Society Proceedings, vol. 4, 1891-93, p. 286-293.
3. American Journal of Science, ser. 3, vol. 32 (132), 1886, p. 310-311.
4. Lapidary Journal, vol. 10, 1956, p. 50.
5. American Journal of Science, ser. 3, vol. 47 (147), 1894, p. 305-309.
6. Rocks and Minerals, vol. 41, 1966, p. 731-732.
7. American Mineralogist, vol. 50, 1965, p. 1204.
8. U. S. Geological Survey Bulletin 1114, 1961, p. 17, 47, 49, 95, 99, 104, 116, 155, 160, 212, 218, 275, 309, 355.

MAPS

Topographic maps: Poncha Springs (1956).
National forests: San Isabel.

Turret District

In the Turret district, northwest of Salida, are a number of pegmatites containing beryl, garnet, columbite-tantalite, and other interesting minerals, as well as an abundance of the more common pegmatite minerals. One of these deposits, consisting of the Homestake mine, is of enormous size. The most noted locality – having yielded true gem sapphire, hessonite garnet, attractive sagenite, and superb epidote – is the Calumet iron mine.

The logs to the Turret localities from downtown Salida are as follows:

0.0 Intersection of Main Street and Colo. 291. Go north on Colo. 291 toward U. S. 285. Cross Denver and Rio Grande Western Railroad.

1.0 Turn right onto Spiral Drive (not marked in 1971).

1.1 Cross bridge over Arkansas River.

1.2 Turn right onto Chaffee County 180.

1.4 Keep left across railroad at refinery.

1.6 Keep ahead on Chaffee County 180. (Road to right goes up Tenderfoot Hill on Spiral Drive.)

5.3 Enter San Isabel National Forest through cattle guard.

8.1 Keep ahead on Chaffee County 180. (Right road goes 2 miles to Federal granite quarry.)

9.1 Fork left onto Chaffee County 190. (Co. 180 goes right to Whitehorn.)

10.7 Railroad Gulch. Junction of Chaffee County 190 (to Turret) and Co. 31 (to Calumet mine).

The Rock King mine is situated on the left side of this junction. It has been an intermittent producer of beryl from three parallel pegmatite dikes in quartzite. The beryl occurs in crystals as much as 8 feet long and 1 foot across. A little columbite-tantalite is present. Muscovite and biotite mica, microcline and albite feldspar, massive white quartz, and graphic granite make up the bulk of the pegmatite, as described by John B. Hanley, E. William Heinrich, and Lincoln R. Page (Ref. 1).

Continuing log:

10.7 Junction of Chaffee County 190 and Co. 31. Go on Co. 31 toward Calumet mine.

11.0 Homestake mine.

This huge quarry consists of several openings in pegmatite and a complex network of roads for the removal of material. The country rock is gray quartzite and chlorite gneiss. Albite feldspar is the chief mineral in the deposit, and there is a substantial amount of microcline feldspar, muscovite mica, and quartz. Fluorapatite is the main accessory mineral. This property, often referred to as the Soda Spar Quarry, has been vastly enlarged since it was described by Hanley, Heinrich, and Page (Ref. 2), and was recently the largest feldspar operation in Colorado, owned by M & S Inc. (Ralph Magnuson), of Salida. It was closed after a period of declining production since 1960. It was fenced in 1962 for protection and quit mining in 1963.

Continuing log:
11.0 Pass Homestake mine.
11.9 Calumet iron mine. Mine timbers on right. Largest dumps below road on left.

This is one of the most interesting gem-collecting localities in Colorado. From it have come gem sapphires, hessonite garnets, pretty sagenitic quartz crystals, and some of the best epidote ever found in the West. These minerals were first described in 1887 by Walter B. Smith (Ref. 3).

The abandoned Calumet iron mine is situated on the side of a hill having an altitude of about 9,500 feet. When operated from 1882 to 1900 by the Colorado Fuel and Iron Company, it was served by the Denver and Rio Grande Western Railroad. The huge mine workings extend over ¼ mile along the hillside and an equal distance up the hill. Charles H. Behre, Jr., E. R. Osborn, and E. H. Rainwater, who studied the area in detail, described it (Ref. 4) as a typical contact-metamorphic deposit. A large magnetite vein furnished the iron ore.

The best collecting is on the upper workings. Specimens of epidote and quartz were originally uncovered by workmen who found them lying loose in the dirt while they were stripping the hillside for the entrance of the highest tunnel; and later, they were found in pockets in the rock above this tunnel.

Although not over 1 foot thick, the layer of corundum schist in which the sapphires were found was exposed for about 800 or 900 feet. Its present location remains a mystery, perhaps being concealed by slide rock. Above it is a dark-greenish igneous rock, and beneath it is the garnet-bearing bed of marble. In addition to feldspar, quartz, and muscovite mica, the schist contains sillimanite and graphite (Ref. 5), as well as many tabular, rhombohedral crystals of corundum, constituting one-third of the rock and giving it a banded appearance. These crystals range in diameter up to ¼ inch, although most of them are less than half that size. They are deep blue, some having been found to contain transparent areas and to be of true gem-sapphire quality. The larger crystals are the most badly fractured, although the color is very good. The impure corundum was mined to some extent around 1893 to 1895 and was milled at Idaho Springs and made into wheels by the Colorado Abrasive Company, of Denver.

The marble bed beneath the corundum schist is 6 or 8 feet thick, and it contains garnets of the variety of grossularite known as hessonite or essonite. Most of the garnet is massive, but well-formed crystals up to 1 inch in diameter are present, showing bright and smooth dodecahedron and trapezohedron faces. The best have a beautiful, yellowish-brown color, whereas others are quite black. The garnet was discovered by Nathaniel D. Wanemaker, who also found the aquamarine on Mount Antero. The marble bed, which is composed of brown and white calcite, also contains epidote and gray-green, radiating actinolite.

Crystals of epidote occur embedded in soft limestone, lying loose in cavities in a metamorphosed rock, which overlies a bed of limestone, which in turn is above the magnetite vein. The epidote is opaque or faintly translucent, having a typical, pistachio-green color. Crystals have been found up to 2 inches long; and within the past few years, some magnificent and valuable groups have been collected here.

Attached to epidote are transparent quartz crystals of the sagenite variety, being penetrated by fibrous epidote (formerly called byssolite). The crystals range up to 6 inches in length and have a tapering form without terminations.

Exceptionally fine specimens of diopside – the chief silicate mineral here – are found in large, radiating sheaves at the Calumet mine. Crystals of magnetite, the ore mineral of this locality, are found up to about ½ inch in size. Crystals of augite are 1 inch or more in length. Other minerals, some of which make splendid cabinet specimens, are hematite, pyrite, chalcopyrite, feldspar, brown and green biotite mica, pyrophyllite (?), and wernerite.

Return to junction of Chaffee County 190 and Co. 31. Continuing log:

10.7 Go on Co. 190 toward Turret.

11.1 Park at three-way fork. Walk left about 100 yards along road to Savilla Queen 2 mine. This general pegmatite operation dates from 1951.

Continuing log:

11.1 Take middle road at three-way fork.

11.4 Turn left at fork. Go 100 yards to group of prospects.

Formerly called the Last Chance Spar-Mica Dyke prospect and then the Old Glory claim, this was renamed the Blue Brute mine, claimed by Glen R. Lamberg, of Salida. Hanley, Heinrich, and Page (Ref. 6) have described the presence of small beryl crystals and dark-brown garnet in typical pegmatite feldspar (both potash and plagioclase), quartz, and muscovite.

Continuing log:

11.4 Go on main road toward Turret.

12.2 Keep ahead. (Right track goes to Golden Wonder mine.)

13.6 Turret.

Once a prosperous gold-mining town, Turret had a population of one when the writer first visited it in 1938, increasing to five 8 years later, and fluctuating slightly since then.

The Combination mine is 200 yards south of the road after it forks left at Turret. Blue crystals of beryl up to 1 foot in length have been found in this pegmatite. Crystals of garnet, red and white feldspar, plates of biotite, masses of quartz, and graphic granite constitute the rest of the rock, as described by Hanley, Heinrich, and Page (Ref. 7).

REFERENCES

1. U. S. Geological Survey Professional Paper 227, 1950, p. 27.
2. U. S. Geological Survey Professional Paper 227, 1950, p. 23-24.
3. Colorado Scientific Society Proceedings, vol. 2, 1885-1887, p. 163-166, 176-177.
4. Economic Geology, vol. 31, 1936, p. 781-804.
5. American Mineralogist, vol. 33, 1948, p. 199.
6. U. S. Geological Survey Professional Paper 227, 1950, p. 24-25.
7. U. S. Geological Survey Professional Paper 227, 1950, p. 23.
8. Rocks and Minerals, vol. 22, 1947, p. 16-17.
9. Rocks and Minerals, vol. 23, 1948, p. 496, 552.
10. Mineralogist, vol. 9, 1941, p. 45-46.
11. U. S. Geological Survey Bulletin 269, 1906, p. 55, 141.

12. Gems and Minerals, no. 311, 1963, p. 26.
13. U. S. Geological Survey Bulletin 1114, 1961, p. 16, 112, 116, 133, 147, 159, 174, 180, 210, 236, 271, 275, 292.

MAPS

Topographic maps: Cameron Mountain (1956).
National forests: Pike, San Isabel.

BUENA VISTA TO COLORADO SPRINGS

Most of the long road from Buena Vista to Colorado Springs, which joins U. S. 285 for awhile, crosses South Park, a wide expanse of high ranch land surrounded by mountains. Translucent petrified wood, the finest known in Colorado, has come from various isolated places in South Park, and other kinds of chalcedony are found here. Petrified wood is very abundant at certain horizons in certain localities in South Park. Beautiful almandite garnets of gem quality have been found in the placers since they

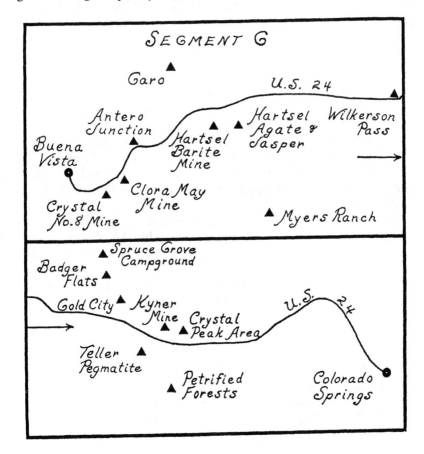

were first described by J. Alden Smith in 1870. A pegmatite in Chaffee County, west of Trout Creek Pass, and a travertine locality in Park County, east of the same pass, are described below. The barite mine at Hartsel, described below, is of exceptional importance, and the blue-agate and radioactive-mineral locality of Garo is of some interest. The Meyers Ranch pegmatite, described below, is reached from Hartsel. Pegmatites and metamorphic deposits near Wilkerson Pass, the beryllium deposits at Badger Flats, and pegmatites near Lake George and in the Tarryall Mountains are described below. The world-famous Crystal Peak locality, in Teller County, described below, is outstanding, as are the petrified forests near Florissant, mentioned below. The highway reaches Manitou Springs and Colorado Springs, in El Paso County, by way of Ute Pass.

Clora May Mine

A new locality (which we have named Jasper Hill) was found in 1969 by Mrs. Pearl on the slopes above the feldspar quarry known as the Clora May mine, situated near U. S. 24-285 west of Trout Creek Pass. Huge boulders, as well as many broken fragments, of red and yellow jasper, jasper conglomerate, variegated chalcedony, and sparkling patches of drusy quartz, lie strewn on the ground to the left of the road that winds beyond the mine. The jasperized area gradually gives way to volcanic rock. Look further for more jasper!

At least two and possibly four rare-earth minerals occur in the Clora May. The Yard mine, when owned by the M and S Company, of Denver, had 1,000 tons or more of feldspar in reserve a few years ago, but had become inactive. A new claim, called the Mina Blanca, was made in 1961. About 600 pounds of bismuthinite and secondary bismuth minerals (especially bismutite) were produced here. The main minerals of interest are now euxenite, allanite, and rose quartz, the last of which is abundant, some of it being attractively banded.

The log to this mine from the west is as follows:

0.0 North corner of triangle at junction of U. S. 24 and U. S. 285. Go east on U. S. 24-285 past Johnson Valley.

5.9 Crystal No. 8 mine situated across canyon on right of highway.

108

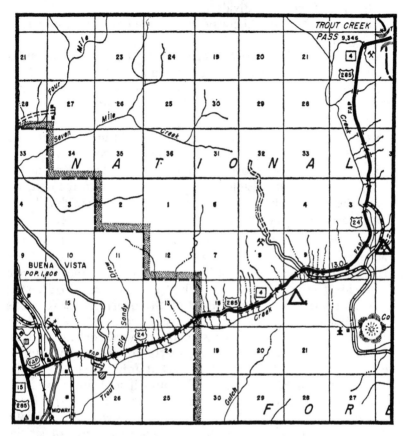

6.4 Turn right onto Chaffee County 53 toward Bassam Park.

6.5 Keep ahead at junction. (Right road goes to Crystal No. 8 mine, included in previous edition of this book; now fenced off.)

8.0 Keep ahead toward Trump. (Right fork goes 2 miles to Sedge peat bog, source of peat for fertilizer use.)

8.4 Park on road below hill on right. Two branches of an old truck road are visible at the right, going up to the mine on the north side of the hill; can be driven when dry, lower road best; road above mine swings left to Jasper Hill.

The log to this mine from the east is as follows:

0.0 South corner of paved triangle at Antero Junction. Go west on U. S. 24-285.

0.8 Trout Creek Pass divide.

5.0 Turn left across highway just before crossing Trout Creek bridge.

5.5 Road to white quartz ledge on left. The quartz protrudes from dark quartz-mica schist, which is the country rock. Red feldspar was presumably mined here, but not more than a few tons could have been taken out, and very little is left.

5.8 Left road (Mushroom Gulch) forks right in .4 mile around Jasper Hill to top of Clora May mine. Or continue.

5.9 Park on road below hill at left. Two branches of an old truck road are visible at the left, going up to the mine on the north side of the hill; can be driven when dry, lower road best; road above mine swings left to Jasper Hill.

The Clora May quarry stands at the top of the hill. It consists of an opencut over 100 feet long, 40 feet wide, and 30 feet deep, excavated in pegmatite. Subordinate cuts surround the main one on the side facing the road. Abundant white quartz float lies on the steep slopes of the hill. Some of the quartz is smoky. The country rock is a dark quartz-mica schist.

Euxenite, which is a brownish-black mineral containing uranium and several elements of the rare-earth group, was reported as rare by E. William Heinrich (Ref. 1), but John B. Hanley, Heinrich, and Lincoln R. Page (Ref. 2) found it as being present in masses 8 to 10 inches across, showing occasional crystal faces. Even the smaller pieces, however, are not now easy to locate. Bladed crystals of allanite, another rare-earth mineral of similar appearance, have likewise been found here. Gadolinite and xenotime are of scarce occurrence.

Pink microline feldspar was the chief mineral mined at the Clora May, even though the bismuth ores were a substantial by-product. Some of the feldspar is darker, containing more iron. Red albite feldspar occurs in parallel crystals, somewhat altered. Graphic granite, an intimate intergrowth of feldspar and quartz, is fairly abundant. Large pieces of black hornblende were common. Considerable amounts of biotite mica, some in books of crystals 2 feet in diameter, occur at the outer edges of the core zone. Some poorly developed garnet crystals are available. Mrs. Pearl in 1969 found considerable black tourmaline, not previously described here. Green and purple fluorite occur here. Apatite, calcite, limonite, and sericite are not specimen material.

REFERENCES

1. American Mineralogist, vol. 33, 1948, p. 64-75.
2. U. S. Geological Survey Professional Paper 227, 1950, p. 22.
3. U. S. Geological Survey Bulletin 1114, 1961, p. 17, 74-75, 135, 154.

MAPS

Topographic maps: Antero Reservoir (two scales; 1956, 1959).
National forests: Pike, San Isabel.

Antero Junction

Reported to the author by Dr. Don B. Gould, of Denver, who has described it in print (Ref. 1), a hot-spring deposit of travertine is located a short distance from Antero Junction. Lying loose among the many surface boulders of travertine are specimens of jasper and agate.

The log to this locality from Antero Junction is as follows:

0.0 North corner of paved road triangle. Go north on U. S. 285 toward Fairplay.

0.7 Stop on crest of hill.

The slopes of both sides of the road are littered with weathered fragments of limestone and boulders of travertine. The travertine is gray and porous, as is typical of this material. It contains bubbly growths and is marked in places by tiny crystals of calcite, presumably deposited in openings by later mineral-bearing water. Orange, yellow, and reddish-brown lichens cover many of the pieces of travertine.

The agate includes some of a mossy variety. Along with common, brown jasper are pieces of cherry-colored material up to 3 inches long. Occasional fragments of jasper bear evidence of Indian flaking, though no pronounced arrowheads have been noted. Some of the material shows a bright-yellow fluorescence under long-wave ultraviolet light and a slight effect under short-wave radiation.

The travertine of this deposit, now weathered and overgrown

111

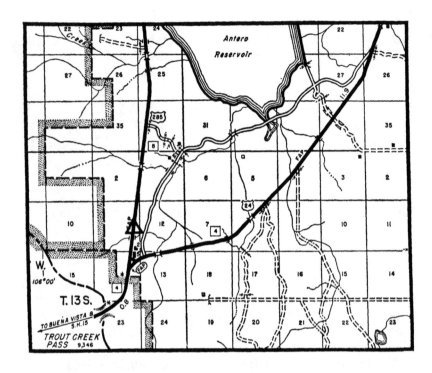

with grass, represents precipitation by a hot spring of Tertiary or later age, unrelated to any existing spring but probably related to the lava flows and other volcanic activity of an earlier geologic origin. The travertine is in terraces, which cover 20 acres or more and are similar to those of Yellowstone National Park. The underlying rock is the Maroon Formation, of Pennsylvanian-Permian age.

REFERENCES

1. Iowa Academy of Sciences Proceedings, vol. 41, 1934, p. 241.
2. Geological Society of America Memoir 33, 1949.

MAPS

Topographic maps: Antero Reservoir (two scales: 1956, 1959). National forests: Pike, San Isabel.

Hartsel Barite Mine

Large and splendid groups of blue barite crystals embedded in clay in a limestone bed are abundant 2 miles southwest of Hartsel.

This is for the collector one of the major mineral localities in Colorado, yielding with little difficulty many handsome specimens in single crystals, in clusters, and in massive form.

The log to this deposit from Hartsel is as follows:

0.0 Junction of U. S. 24 and Colo. 9. Go west on U. S. 24 toward Buena Vista.

0.9 Keep left on U. S. 24 at west junction with Colo. 9.

1.6 Turn left opposite ranch road, go through gate and close it.

2.6 Fork right and follow tracks up slope to join better road coming from right.

2.8 Fork right and bear right.

3.2 Park near workings.

The deposits are in an extensive area of diggings and dumps on the slope of a hill that provides a beautiful view of South Park and

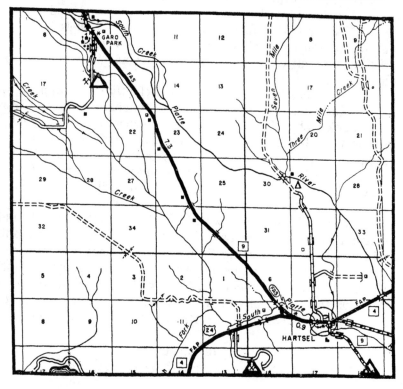

the Park Range. Large amounts of barite lie loose on the dumps. Mining claims were first staked here in 1932 and have been developed intermittently since. An early opening goes underground and is probably dangerous to enter. Besides the shallow pits dug with hand tools, deeper and broader excavations have been opened by bulldozers.

The barite occurs as porous aggregates of crystals mixed with iron-stained clay derived from the original limestone. This gray, brown-stained limestone is a member of the Maroon formation of Pennsylvanian-Permian age, which consists largely of siltstone. The barite is found in layers 6 inches to 3 feet thick and also in vertical veins 1 to 2 feet thick, which cut across the limestone beds. Combinations of the two types also exist. Calcium carbonate encrusts the surface of some of the crystals. Iron oxide is stuck to the bottom of some of the groups.

The best barite is found in tabular crystals up to 5 inches long and 2 inches high. Many of them are well developed and show good basal and prism faces typical of barite, as well as an occasional uncommon form. Many groups of crystals represent the terminal parts of huge masses of barite reaching more than 1 foot in length.

The crystals of barite are striking chiefly on account of their blue color, similar to the barite near Stoneham, described later in this book. The Hartsel crystals are generally coarser and less transparent, but some are delicate, sharp, and clear. The color is irregularly distributed, disappearing in the center of the larger masses. The blue color may have resulted from radioactivity, as suggested by Arthur L. Howland, who described the geology of this locality (Ref. 1). His opinion is based on the proximity of the Hartsel Hot Springs, the water of which is rather highly radioactive, and the Garo deposit of radioactive vanadium and uranium minerals about 8 miles north. Radiation is known to disturb the structure of various minerals, producing a blue color in fluorite, halite, barite, and others. The ultraviolet radiation of the sun deepens the color of the Hartsel barite, making the exposed specimens on the dump more attractive than those still buried.

REFERENCES

1. American Mineralogist, vol. 21, 1936, p. 584-588.

2. Colorado School of Mines Quarterly, vol. 44, no. 2, 1949, p. 34.

MAPS

Topographic maps: Hartsel (1956).
National forests: Pike.

Garo

This interesting locality – Garo or Garo Park – between Hartsel and Fairplay is a former uranium deposit now attractive for its blue agate and other varieties of chalcedony, which lie strewn across the field on which heaps of radioactive ore used to lie. Drusy quartz is found here in large pieces. The agate is partly mossy, and much of the chalcedony is fluorescent.

The red sandstone belongs to the Maroon Formation, of Pennsylvanian and Permian age. It is locally rich in carnotite, now seen here only in meager amounts. The other uranium-bearing mineral, an uncommon one, is calciovolborthite. Secondary copper sulfides and carbonates accompany them. A Geiger counter will be useful in searching for these minerals, but the blue agate is almost certain to be the reason for collectors to stop here. The original description (labeled Garos) was by Herman Fleck (Ref. 1). The ore was trucked to Rifle for processing.

This locality can be reached from Garo (not marked) by the following log:

0.0 Turn south onto Park County 24 just north of Garo.
1.0 Locality on left. Field cut up by pits and trenches dug while mining uranium.

REFERENCES

1. Mining World, vol. 30, 1909, p. 597.
2. U. S. Geological Survey Circular 220, 1952, p. 27, 28.
3. U. S. Geological Survey Circular 215, 1953, p. 3.
4. U. S. Geological Survey Bulletin 1114, 1961, p. 26, 85, 90.

MAPS

Topographic maps: Garo (1956).
National forests: Pike.

Meyers Ranch

Among the best and most abundant gem rose quartz in Colorado occurs in the Meyers Ranch pegmatite, which is situated in Park County 20 miles south of Hartsel on Colo. 9. In addition, there are a number of other interesting minerals, which include cordierite, bright-blue to olive-yellow or golden crystals of beryl as much as 10 inches across, columbite-tantalite in crystals up to 6 inches long, black tourmaline, and brown garnet. Although cordierite is well known in this Guffey-Micanite district, it had not been described from this mine until identified in 1964 from specimens found by Mrs. Pearl. Cream-colored crystals of microcline feldspar, averaging 6 feet in size, gray-green to "rum-colored" books of muscovite mica up to 8 inches across, albite plagioclase feldspar, biotite mica, fluorapatite, bismutite, and beyerite are the other minerals present. Microcline and muscovite make up most of the deposit apart from the quartz, which (as two separate pods) constitutes the core of the pegmatite. This lens-shaped body is enclosed in red granite gneiss of Precambrian age. Lava rock is found nearby.

This mine was discovered in 1908. It was owned by Alonzo MacDonald and later by Jesse E. Meyers and has been worked on several occasions, producing feldspar, scrap mica, beryl, and columbite-tantalite. The workings consist of various opencuts drifts, and prospect pits. The mine was described by John B. Hanley, E. William Heinrich, and Lincoln R. Page (Ref. 1). Heinrich (Ref. 2) described the occurrence here of bismutite and beyerite, the latter a rare mineral first analyzed from specimens found at this place and near the Royal Gorge.

The Meyers Ranch pegmatite can be reached from Hartsel according to the following log:

0.0 Junction of U. S. 24 and Colo. 9. Go on Colo. 9 toward Canon City.
19.4 Pass cluster of log cabins on both sides of road.
19.5 Cross cattle guard, turn right through gate and close it, ford Currant Creek, and bear right to mine.
20.0 Park at mine.

The Colorado Bureau of Mines records a mine under the name Meyers Ranch quarry as being southwest of Guffey and worked by Rock Products, Inc., of Florence.

REFERENCES

1. U. S. Geological Survey Professional Paper 227, 1950, p. 105-107.
2. American Mineralogist, vol. 32, 1947, p. 660-669.
3. U. S. Geological Survey Bulletin 1114, p. 20, 237.

Wilkerson Pass

Black tourmaline can be found on a number of scattered places near the camp at Wilkerson Pass (9,525 feet) on U. S. 24. A variety of minerals is available in both directions from the pass, as indicated in the following short logs that start here. In addition to those mentioned are magnetite, garnet, scheelite, sphalerite, and galena.

0.0 Top of Wilkerson Pass. Go west toward Hartsel.
2.4 Turn right onto gravel road. Deposits scattered on open ground ahead.

The pegmatites contain tourmaline, beryl, quartz, feldspar, and biotite and muscovite mica.

0.0 Top of Wilkerson Pass on U. S. 24. Go east toward Lake George.
0.5 Turn left onto mine road.
0.6 Park. Mine dumps above, on left, are reached by moderate climb.

Various metamorphic minerals can be collected here. Epidote is prominent, and books of biotite mica are common. Green stains indicate the presence of copper ore. The country rock is mostly gneiss and schist of no specimen interest.

0.0 Top of Wilkerson Pass. Go east toward Lake George.
2.0 Turn left onto mine road. Abandoned St. Joe tunnel is ahead. Fork right across open ground.
0.0 Top of Wilkerson Pass. Go east toward Lake George.
2.0 Turn left as logged above. Fork right across open ground.
2.5 Curve left to foot of hill. Mine is on side of hill.

The shafts are dangerous.

Chalcopyrite, bornite, pyrite, malachite, and azurite were seen here.

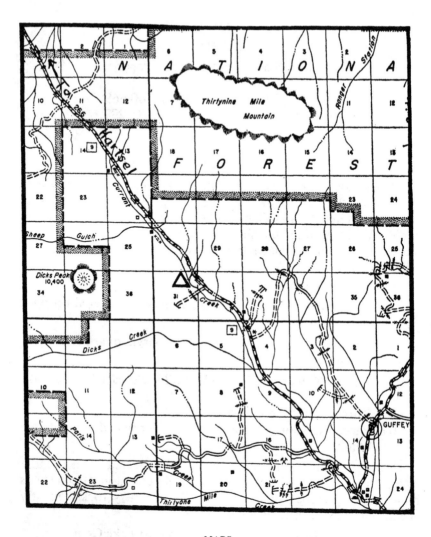

MAPS

Topographic maps: Glentivar (1956), Tarryall (1956).
National forests: Pike.

Badger Flats

The largest beryl mine in the United States during several recent years (1959 and since) and a number of smaller mines and diggings are situated at Badger Flats, near Wilkerson Pass. The

118

names Tarryall and Lake George have also been used for this area. The Boomer mine is operated by the U. S. Beryllium Corporation, of Pueblo. Milling of ore from elsewhere is carried on here as well. In addition to its size in terms of both production and reserves, the Boomer mine is noteworthy because the mineralization is in veins, pipes, pods, and complex, irregular bodies enclosed in greisen, rather than pegmatite, from which virtually all the world's beryl had previously been taken. As specimens, therefore, the minerals that occur here are small and not especially attractive. Beryl (much altered to sericite mica), bertrandite, and euclase contain beryllium; the other minerals include topaz, fluorite, siderite, galena, pyrite, sphalerite, molybdenite, wolframite, chalcopyrite, arsenopyrite, cassiterite, pitchblende, muscovite mica, and quartz.

The log to this locality from Lake George is as follows:

0.0 Junction of U. S. 24 and Eleven Mile Canyon road. Go on U. S. 24 toward Hartsel.
0.5 Cross South Platte River.
1.2 Pass Park County 77 on right. (Road to right goes to Hyner mine and Spruce Grove Camp Ground, described below.)
5.8 Turn right onto gravel road.
6.1 Fork right.
6.4 Turn right through campground (federal fee for camping).
6.6 Go through gate and close it. Pass smaller mines and diggings on both sides.
9.6 Go through gate and close it.
9.9 Boomer mine ahead. Blue Jay mine at right. Others in the vicinity are the J.S., Happy Thought, Tennessee, Mary Lee, Little John, and Redskin.

Scheelite deposits with a different kind of mineralization occur elsewhere in this area, as do vein deposits of fluorite, barite, and precious and base metals, but these are not further described here, not having collecting value.

REFERENCES

1. U. S. Geological Survey Bulletin 1114, 1961, p. 26, 70, 73-74, 92, 335, 357.
2. U. S. Geological Survey Professional Paper 400-B, 1960, p. B71-B74.

3. American Mineralogist, vol. 46, 1961, p. 1505-1508.
4. U. S. Geological Survey Professional Paper 501-A, 1964, p. A9.
5. U. S. Bureau of Mines Report of Investigations 6828, 1966.
6. U. S. Geological Survey Professional Paper 550-C, 1966, p. C138-C147.
7. U. S. Geological Survey Circular 597, 1968.
8. U. S. Geological Survey Professional Paper 608-A, 1969.

MAPS

Topographic maps: Tarryall (1956).
National forests: Pike.

Gold City

Good crystals of green idocrase and bright crystals of orange to brown garnet are found with quartz and potash feldspar in a metamorphic deposit enclosed in Pikes Peak granite near the ghost town of Gold City. Other minerals occur nearby. This locality was made known to the writer by the discoverer, Timothy C. Anglund, who recovered specimens of the idocrase for sale. He had requested that it not be described during his lifetime.

The log from Lake George is as follows:

0.0 Junction of U. S. 24 and Eleven Mile Canyon road. Go on U. S. 24 toward Hartsel.
0.5 Cross South Platte River.
1.2 Pass Park County 77 on right. (Road to right goes to Kyner Mine and Spruce Grove Camp Ground, described below).
2.9 Mine dumps on hillside on left.

Above the dumps is a small quarry containing pink feldspar, white quartz, graphic granite, long blades of biotite mica, and a little muscovite mica. The pegmatite is in Precambrian gneiss.

Farther along, the hillside is cut by another pegmatite.

Directly across the highway and much higher on the hill is a mine that has yielded garnet and idocrase. Specimens can still be found on the dump, according to Ron A. Timroth.

Continuing log:

3.3 Pass brown farmhouse on right.
4.1 Pass old farm buildings on left.

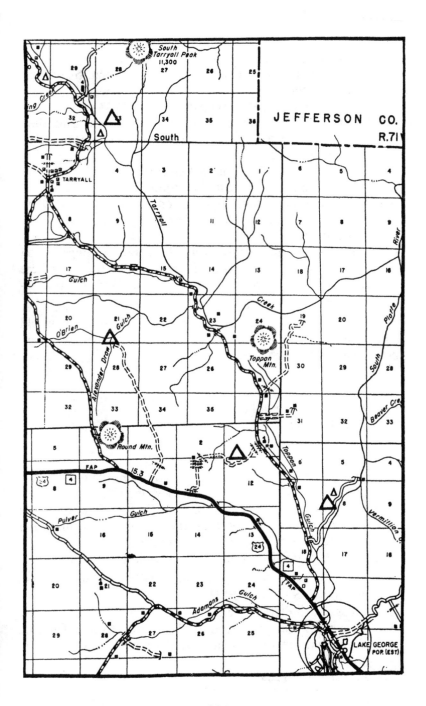

4.9 Pass remains of old farm buildings on left.

5.0 Turn right through fence gate onto dirt road. Sign: Vernon Thatcher.

5.7 Go through and close farm gate.

5.9 Go through and close farm gate (if closed) and then another similar gate, staying to right of buildings. Keep to right along fence.

6.0 Turn right and follow land between fences.

6.1 Go through gate and close it (if found closed).

6.2 Fork right onto faint road, keeping to right of gulley. Ruins of old mill on hill to left are a relic of Gold City ghost town.

6.5 Park in opening on left. Several separate diggings are on the side of the hill to the left, reaching to the top where the largest workings are. Quartz and feldspar float is abundant on the slopes.

Across the valley a little farther back is the shaft of an old molybdenite mine. At the top of the hill and farther along are found specimens of hematite.

MAPS

Topographic maps: Tarryall (1956).
National forests: Pike.

Kyner Mine

Quartz crystals are readily secured from pegmatite in the Pikes Peak granite in several places close together along the South Platte River, near the resort town of Lake George, in Park County. Some of the productive rock lies loose on the surface of the hills, and the rest constitutes bedrock still fixed in place.

The log to this locality from Lake George is as follows:

0.0 Junction of U. S. 24 and Eleven Mile Canyon road. Go west on U. S. 24 toward Hartsel.

0.5 Cross South Platte River.

1.2 Turn right onto Park County 77 (gravel) toward Tarryall.

2.6 Turn right onto dirt road toward Happy Meadows Campground.

3.1 Park ahead on right.

To the left, above the road, are a number of shallow diggings, representing a search for gold and feldspar and other pegmatite minerals. The rock is typical Pikes Peak granite, showing the characteristic, spheroidal weathering, breaking into rounded boulders somewhat buried in thick gravel. The pegmatite is partly massive feldspar, partly white quartz, and partly graphic granite. Drusy, white quartz coarsens to larger crystals having a similar appearance. Good quartz crystals a couple of inches long were found here by Estle Johnson, of Colorado Springs, who first made the author acquainted with this locality. Mrs. Pearl picked up crystals as much as 2 inches wide on a previous visit here. Toward the top of the shallow draw are found blue cleavage pieces of fluorite.

Continuing log:

3.1 Parking space mentioned above. Continue.

3.4 Park on left. Abandoned workings and dumps of Kyner mine. These can also be reached from the previous spot by walking down from the crest of the ridge. Happy Meadows Campground on road ahead.

Similar specimens are obtainable here, with perhaps less quartz and more fluorite. This land has recently been withdrawn by the U. S. Forest Service from mineral entry as a mining claim. It was formerly worked by the Lake George Fluorspar Company. The widest vein was 20 inches, along which an open cut was mined to a distance of 110 feet, according to George O. Argall, Jr. (Ref. 1).

Other, similar localities are situated in this area and may be reached by following the outcrops and old diggings. At 5.0 miles is the entrance to Stevens Ranch, a classic locality for amazonstone; smoky quartz, goethite, and pseudomorphs of hematite after siderite were also found here in splendid quality. This locality is now closed to collectors.

REFERENCES

1. Colorado School of Mines Quarterly, vol. 44, no. 2, 1949, p. 190.

MAPS

Topographic maps: Hackett Mountain (1956).
National forests: Pike.

Spruce Grove Camp Ground

The discovery, by Edwin W. Over, Jr., in November 1962, of a blue crystal weighing more than 1 pound, brought to attention again the merits of the Tarryall Mountains as a source of topaz. These mountains extend southeastward along the northeastern edge of South Park, in Park County. The Lone Lode pegmatite contains large, rough crystals of colorless to blue topaz, as reported by John W. Adams to Edwin B. Eckel (Ref. 1). A man named Crabtree has been credited with the first discoveries in the Tarryalls.

The best known locality in this area is Spruce Grove (or Spruce) Camp Ground. The earliest work here seems to have been done by George Reeser about 1909. In 1929, Mr. Reeser, then about 80 years of age, led Willard W. Wulff and George M. White, among the leading mineral collectors of Colorado Springs, to his old diggings, where they obtained many more specimens and expanded the area by finding new spots farther up the ridge. Mr. Wulff described his experience in an article written in 1939 (Ref. 2). Others have collected here since, and both the Colorado Mineral Society and the Colorado Springs Mineralogical Society have conducted organized trips with varying degrees of success. The latter club maintains a mining claim.

The log to this locality from Lake George is as follows:

0.0 Junction of U. S. 24 and Eleven Mile Canyon Road. Go west on U. S. 24 toward Hartsel.
0.5 Cross South Platte River.
1.2 Turn right onto Park County 77 (gravel) toward Tarryall.
2.6 Road to right goes to Kyner mine, described above.
13.7 Keep ahead and to right through Tarryall settlement.
15.3 Turn right at triangular junction, toward Spruce Grove Camp Ground.
15.5 Spruce Grove Camp Ground (federal fee or permit for camping). Cross creek on foot bridge, follow trail directly uphill through aspen grove; turn right on intersecting trail just below mining prospect (avoiding blue-blazed trail nearby), and continue generally southeast to claim; total distance .8 mile.

The topaz occurs in pockets in pegmatite in the Pikes Peak granite, but is recovered almost entirely from the loose rock, which has weathered to a coarse gravel. The crystals of topaz range up to about 1 inch in length; they are colorless or pale blue, but the cleavage cracks are stained red by iron. The prism faces are striated lengthwise and many are etched; they are terminated at one end only, the other being a cleavage surface. Associated with the topaz are smoky quartz, rock crystal, orthoclase feldspar, biotite and muscovite mica, and hematite.

REFERENCES

1. U. S. Geological Survey Bulletin 1114, 1961, p. 335.
2. Rocks and Minerals, vol. 9, 1934, p. 45-47.
3. Gems and Minerals, no. 309, 1963, p. 16-19.
4. Lapidary Journal, vol. 22, 1968, p. 138-143.

MAPS

Topographic maps: McCurdy Mountain (1956).
National forests: Pike.

Teller Pegmatite

Edwin W. Over, Jr., mined gadolinite, yttrofluorite (yttrium fluorite), monazite, and xenotime several years ago in this pegmatite body near Lake George. The deposit is about 100 feet long and 25 feet high. Several pits have been excavated and are apt to contain water, a possible hazard.

A fairly good road leads through a scenic canyon ahead, with picnic and fishing areas, to Eleven Mile Canyon Reservoir.

The log to this locality from Lake George is as follows:

0.0 Junction of U. S. 24 and Eleven Mile Canyon road. Go south on gravel road past post office toward Eleven Mile Canyon Reservoir.
0.9 Keep right at junction.
1.1 Teller pegmatite on left.

Allanite and molybdenite have also been described from here (Ref. 1); bastnasite is a possibility; the author found astrophyllite; and biotite mica is conspicuous, as are pink microcline feldspar and white quartz.

REFERENCES

1. American Mineralogist, vol. 43, 1958, p. 991-994.
2. U. S. Geological Survey Bulletin 1114, 1961, p. 27, 37, 67, 153, 155, 232, 360.

MAPS

Topographic maps: Florissant (1959), Lake George (1956). National forests: Pike.

Crystal Peak Area

North of Florissant is the Crystal Peak area, the finest producing locality in the world for amazonstone. Magnificent smoky quartz, topaz, phenakite, and amethyst are other gem crystals taken from here. Fluorite, goethite, columbite, cassiterite, and minor iron minerals also occur near Crystal Peak. Hundreds of diggings have been worked intermittently since about 1865 and account for a large share of the Pikes Peak production, which is estimated at one time to have been over $1,000 annually in cut gems, and probably many dollars more in crystals. Crystals of smoky quartz and amazonstone equal in size and quality to any found in the past have been taken from this area in recent years by Raymond F. Ziegler and George W. Fisher, of Colorado Springs, C. W. Hayward, of Denver, and other collectors.

A half-smoky-quartz and half-white-quartz crystal weighing 135 pounds was found here (labeled Lake George) in 1964 by Clarence Coil, of Colorado Springs. Except an early-day "lost" crystal said by Orville A. Reese to have weighed more than 400 pounds, this seems to be the largest crystal ever reported from Colorado. In the same pocket, which yielded half a ton of specimens, was found a 36-pound crystal (now in Waynesburg College, Pennsylvania), one of 32 pounds, and one of 26 pounds.

Early in the 1870's, Dr. A. E. Foote, Philadelphia mineral dealer, employed as many as 19 miners and purchased specimens from others, during the most productive era that this locality has ever seen. George Reeser, a Manitou Springs mineral dealer, and Arsene Thiebaud, a settler, uncovered good material before the end of the 19th Century. In the next decade, J. D. Endicott, of Canon City, opened a number of pockets. Then, Albert B. Whitmore established the Crystal Peak Gem Company and sold specimens to visiting collectors and others. With a declining yield, the work at his homestead property, known as the Gem Mines, ceased. The early history of this locality has been interestingly written by George M. White, of Colorado Springs (Ref. 1).

After Mr. Whitmore's death in 1943 at the age of 89, Alexander C. Kirk, former ambassador to Italy, and others acquired ownership of this land and prevented active collecting except by lease. On land owned by Lynn Jeffers, the Lucky Seven Mineral Company operated the Crystal Gem Mines as a tourist concession, charging a fee for a guided trip and all-day collecting rights. In 1964, the area was bought for subdivision, closing these workings. The surrounding region, however, is similarly mineralized and countless unopened pockets should yield specimens for ages to come. This will now have to be done on Pike National Forest land. Mr. Hayward turned over to the Colorado Mineral Society his productive claims adjacent to the Crystal Gem Mines.

Most of the work has been done with small blasts in advantageous places, but many specimens throughout the district have been taken from the rotten rock with pick and shovel. Many pockets are found by following quartz stringers exposed in new road cuts and after heavy rains. Mr. Hayward used a bulldozer successfully. Some of the cavities are arranged systematically, while others are irregularly distributed. They are of various sizes, mostly ranging from 2 by 2 to 4 by 6 feet; the latter size pocket may contain 500 to 1,000 crystals, as reported by Jerome W. Hurianek (Ref. 2), who has done considerable work in the area. The largest cavity known was 15 by 15 by 6 feet and was found and excavated by a Mr. Copeland in 1909 or 1910; it produced about $3,500 worth of smoky quartz, amazonstone, and fluorite crystals.

The Crystal Peak area has its southern limit about 2 miles north of Florissant, in the west-central part of Teller County. It constitutes a rectangle of about 3 square miles, centering in Crystal Peak, the highest of a series of granite knobs, formerly known as the Crystal Beds. Crystal Peak itself was called Topaz Butte in the older literature, and locally also Cheop's Pyramid and the Sore Thumb. The knobs being about 3 miles north of Florissant and extend northwest in a broken ridge for about 3 miles. Little Crystal Peak is about 1/5 mile north of Crystal Peak, Deer Mountain is about 1/2 mile northwest, and Sheeps Head is about

127

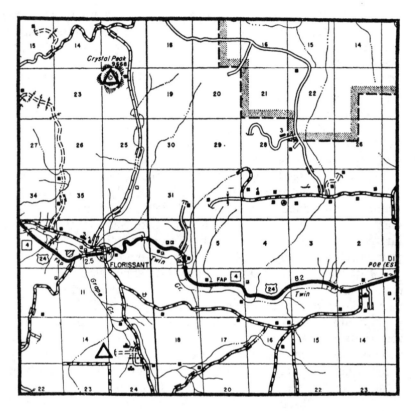

2/3 mile northwest.

The Pikes Peak granite in the Crystal Peak area is commonly coarser and more porphyritic than in the other gem deposits of the Pikes Peak region, and more decomposed and disintegrated than in any other except perhaps at Devils Head, described later. It grades imperceptibly in places into coarse granite and then into aplite. Pegmatite bodies that widen and narrow successively penetrate the badly weathered rock. Inward from the granite, the following zoning is seen: graphic granite, then massive feldspar, quartz, and mica, and finally cavities lined with crystals. The crystals are embedded in moist clay, which usually is easily washed off. The crystals were of course formed in the cavities, singly and in groups, but weathering of the rock has detached many of them, and these are now found broken and loose in the soil. Some are picked up in the gullies as much as 1 mile or more from their place of origin, from where they have been washed. Owing to its perfect cleavage, topaz, however, does not travel intact far from its source.

The crystals of amazonstone range in length up to a recorded 18 inches. Some of the crystals have a clay coating, and some have a capping of white feldspar, apparently the result of secondary growth. Carlsbad, Baveno, and Manebach twins are numerous.

Quartz is the most common mineral in the Crystal Peak deposits, occurring both massive and in crystals, as milky quartz, rock crystal, amethystine quartz, and smoky quartz. Rock crystal is common.

Most of the quartz is the smoky variety, having the deeper color at the top of the usually tapering crystals. Some crystals show phantoms, and others show inclusions of hematite and other minerals. Some are coated with phenakite, goethite, hematite, or limonite, and some have a secondary growth of milky quartz upon them. Mr. Whitmore mined and sold a considerable quantity of smoky quartz, so that the Crystal Peak district ranks higher for this mineral than any other in Colorado. The finest crystal found by him was a twin 12 inches long and 7 inches wide, weighing 52 pounds; it is now in the U. S. National Museum. Among the many superb cut gems originating in this locality is an elliptical brilliant weighing 785 carats, labeled Florissant and displayed in the Isaac Lee collection of the same museum.

Since the first discovery in 1884 (Ref. 3-6), many fine crystals of phenakite have been obtained at Crystal Peak, more than anywhere else in Colorado. They run to smaller sizes than those found at Crystal Park (described later), being in general less than ¾ inch in length. Most of them are colorless, but some may be pale yellow, light gray, or reddish. The phenakite is implanted on other minerals, most commonly on feldspar. Some crystals are enclosed in amazonstone and smoky quartz. Several generations of crystals may be present, as indicated by their different positions relative to other original minerals and their alteration products.

The first of many discoveries of topaz in the Crystal Peak area was by a Mr. Transue in 1884. This was the second find of topaz in Colorado and was described by Whitman Cross and W. F. Hillebrand (Ref. 3). During the next couple of years, the production was estimated at $4,000. The first specimen was 3½ inches long and was believed to have been only part of a crystal almost 1 foot in diameter. Because of its greenish hue, it was mistaken for fluorite. Most of the topaz is colorless, but many of the crystals are greenish, and some have a slightly yellowish or bluish tinge. They occur singly, up to 2 inches long, and in groups, and some are doubly terminated, though usually not symmetrically so. Topaz is often implanted on amazonstone, and rarely is phenakite present without topaz. The most likely pockets for the association of topaz and phenakite seem to be about ½ mile northwest of Crystal Peak, and these two minerals are found only on the west side of the mountain, whereas amazonstone and smoky quartz occur all around it.

Next to quartz, microcline feldspar is the most abundant mineral. Amazonstone, as already described, is the usual variety in cavities. White, gray, and flesh-colored microcline crystals are also present, but not with amazonstone except where they exist as a secondary growth upon the latter. Albite, a plagioclase feldspar, occurs in rounded groups of white or brown color on quartz and amazonstone; it also occurs in single crystals, mostly simple but a few twinned. The feldspars have partly weathered to kaolin.

In some pockets, fluorite is present on smoky quartz and amazonstone. It exists as colorless, green, purple, or blue crystals in combinations of cubes and octahedrons, and most of them seem to be partly dissolved and etched. Groups and cleavage masses have been found weighing over 20 pounds.

Goethite is one of the typical Colorado minerals of the Pikes Peak region that should be familiar to every collector. It occurs at Crystal Peak in mammilliary, radiating crystals and tabular shapes. Pseudomorphs of goethite after siderite are much sought after, as is the onegite variety of quartz, which encloses goethite.

Masses and coatings of hematite are common. Columbite has also been found in a few good crystals on amazonstone. Cassiterite is known here in simple and twinned crystals. Allanite has been reported. Manganite and limonite are present as a crust on other minerals. Leaves of muscovite and biotite mica are common, most of them attached to amazonstone.

REFERENCES

1. Rocks and Minerals, vol. 10, 1935, p. 184-187.
2. Rocks and Minerals, vol. 13, 1938, p. 329.
3. U. S. Geological Survey Bulletin 20, 1885, p. 69-70, 71-72.

4. American Journal of Science, ser. 3, vol. 29 (129), 1885, p. 249.
5. American Journal of Science, ser. 3, vol. 32 (132), 1886, p. 210-211.
6. American Journal of Science, ser. 3, vol. 33 (133), 1887, p. 130-132, 134-135.
7. Mineralogist, vol. 9, 1941, p. 283-284, 311-313.
8. Colorado Scientific Society Proceedings, vol. 2, 1886, p. 108-115.
9. American Mineralogist, vol. 20, 1935, p. 326-328.
10. U. S. Geological Survey Folio 7, 1894.
11. Academy of Natural Sciences of Philadelphia Proceedings, vol. 28, 1876, p. 156.
12. American Journal of Science, ser. 3, vol. 26 (126), 1883, p. 484-485.
13. U. S. Geological Survey Bulletin 1114, 1961, p. 28, 92, 141, 142, 253, 280, 335.
14. Lapidary Journal, vol. 20, 1966, p. 982-989.
15. Lapidary Journal, vol. 24, 1971, p. 1497-1503.

MAPS

Topographic maps: Florissant (1959), Lake George (1956).
National forests: Pike.

Petrified Forests

Petrified tree stumps that are among the largest known, and a remarkable group of three standing stumps called the Trio and unequalled anywhere else in the world, are the outstanding features of two petrified forests situated within a few miles of Florissant. The new Florissant Fossil Beds National Monument includes them, together with the fossil beds that contain a wealth of nearly unique plant and animal remains. It is at present (1971) understood that some collecting may be permitted on the surrounding (private) land. Selenite sheets and groups of crystals of gypsum occur between the layers of Florissant shale.

The log to this area from Florissant is as follows:

0.0 Intersection of U. S. 24 and Colo. 143. Go toward Cripple Creek.
1.8 Entrance to Colorado Petrified Forest on right.

131

2.4 Entrance to Pike Petrified Forest on right.

These forests are described in *Exploring Rocks, Minerals, Fossils in Colorado*, by Richard M. Pearl (Sage Books, The Swallow Press, Inc., Chicago, revised edition, 1969).

MAPS

Topographic maps: Florissant (1959), Lake George (1956).
National forests: Pike.

The valley of the Arkansas River – in places open, elsewhere enclosed in narrow canyons, which culminate in the Royal Gorge – is followed downstream by U. S. 50 as it goes from Salida, in Chaffee County, to Canon City, in Fremont County. Along this route are the travertine quarry at Wellsville, the Devils Hole mine, near Texas Creek, and the splendid array of pegmatites in the Royal Gorge district – one of the world's finest collecting localities – all described below. Closer to Canon City are the Cowan travertine deposits, the sedimentary area below the Skyline Drive, the Garden Park dinosaur-bone locality, the Felch Creek geode locality, and the choice agate of Curio Hill, also described in this section of the book. South of Canon City and Curio Hill, into the lovely Wet Mountain Valley, are reached the historic old mining towns of Westcliffe, Silver Cliff, and Rosita. New industrial materials, including perlite, vermiculite, and thorium ores, formerly came from this area. East of Canon City is the Penrose calcite locality described below.

Collectors are welcome to attend the meetings of the Canon City Geology Club, which meets in the Municipal Building on the third Monday of each month.

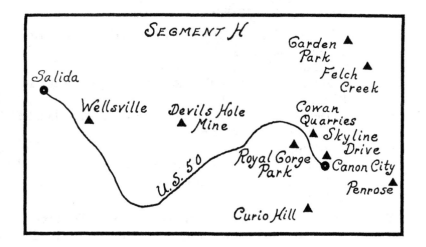

The Municipal Museum in the same building contains fine mineral and fossil exhibits, as well as historical and other displays. This attractive museum is open free from 9:00 to 12:00 and 1:00 to 5:00 daily and from 1:00 to 5:00 on Sunday.

In Pueblo, 39 miles southeast of Canon City, the Pueblo Rockhounds Club, Inc., meets in the McClelland Public Library, 110 East Abriendo Avenue, on the second Friday of each month except July, August, and December.

Wellsville

The largest deposit of travertine in Colorado forms a hill north of the Arkansas River, 3 miles up Taylor Gulch from Wellsville, in Fremont County. It is 1,300 feet long and 200 feet thick, and was estimated by Stephen A. Ionides (Ref. 1) to contain an estimated 50 million cubic feet of travertine. This stone has been used in important buildings throughout the country, including the main stairway in the Department of Commerce Building, in Washington, D. C., the outside of the Sunnyside Mausoleum, in Long Beach, California, and the corridors and interior columns in the City and County Building, in Denver. Blocks as large as 100 tons have been excavated by surface and underground methods at Wellsville. The deposit is now closed after having turned to production of lime for the beet-sugar industry. A nearby quarry now yields limestone for use as a soil conditioner.

The log from downtown Salida is as follows:

0.0 Intersection of U. S. 285 and Colo. 291. Go south on Colo. 291 toward U. S. 50.

1.0 Turn left onto U. S. 50 at junction.

5.9 Turn left onto road to Wellsville. Cross bridge over Arkansas River.

6.6 Cross Denver and Rio Grande Western Railroad and cattle guard. Take middle fork ahead.

6.8 Turn left uphill.

7.0 Next road to right leads to quarry, which is on school land. Road is wide but rutted and may not always be passable in regular car.

The color of the travertine varies from tan to gray. The texture is more porous toward the top. Casts of pine cones and needles have been found, indicating the former presence of vegetation at

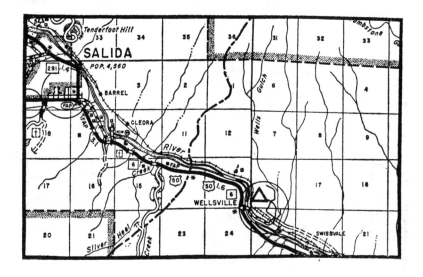

the site of the hot spring that gave rise to the deposit. An abundance of attractive calcite crystals occurs along the rock surfaces and in cavities.

REFERENCES

1. Colorado Mining Association Mining Yearbook, 1936, p. 16-18.
2. Colorado School of Mines Quarterly, vol. 44, no. 2, 1949, p. 459-464.

MAPS

Topographic maps: Howard (1959).
National forests: Rio Grande, San Isabel.

Devils Hole Mine

A substantial source of gem-quality rose quartz, and once reported to contain some gem aquamarine, the Devils Hole mine used to be primarily a producer of beryl, feldspar, and mica. It is now being mined for rose quartz (largely for decorative purposes, the best in Colorado), as well as some beryl and columbite-tantalite. Owing to the dangerous condition of the road while mining is in progress, the new owners, a partnership of two Tezak brothers, require visitors to make arrangements at the Texas Creek

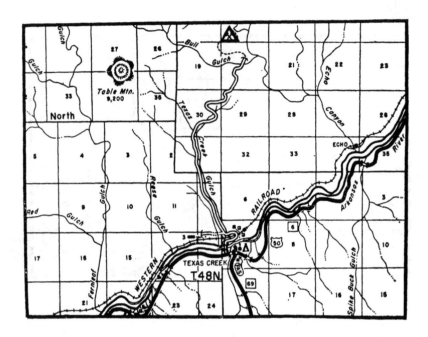

Cafe and Standard Station before entering the property. Ed Tezak, Jr., and his family have been cooperative with collectors; a fee of $2.00 per adult person per day has been charged, but children are free, though not encouraged, for reasons of safety.

The log to the mine from Texas Creek is as follows:

0.0 Junction of U. S. 50 and gravel road at post office and Standard station. Go north on gravel road. Cross bridge over Arkansas River and keep left at fork.

0.2 Turn left and cross Denver and Rio Grande Western Railroad at crossing sign, going through metal gate and close it.

0.5 Keep right at fork.

2.9 Keep right at fork. (Left road may end above the mine; closed after hours 0.5 mile up.) Road follows creek bed for some distance, then becomes very rough as it climbs steeply over a pass, beyond which the mine is seen high on a hill to the north. Road then descends rapidly. On working days, this road is too dangerous to traverse.

5.9 Pass cabins.

6.0 Junction of mine roads. Mine visible on left, about 500

Blue barite crystals from Stoneham. *Ward's Natural Science Establishment.*

Cluster of amphibole (uralite) from the Calumet mine. *Ward's Natural Science Establishment.*

137

The Arkansas River in the Royal Gorge. *Stewart's.*

Cowan travertine quarry near Canon City. *Colorado Advertising and Publicity Department.*

Pikes Peak from the north, showing Glen Cove. *Stewart's.*

Smoky quartz crystals, Crystal Peak. *Ward's Natural Science Establishment.*

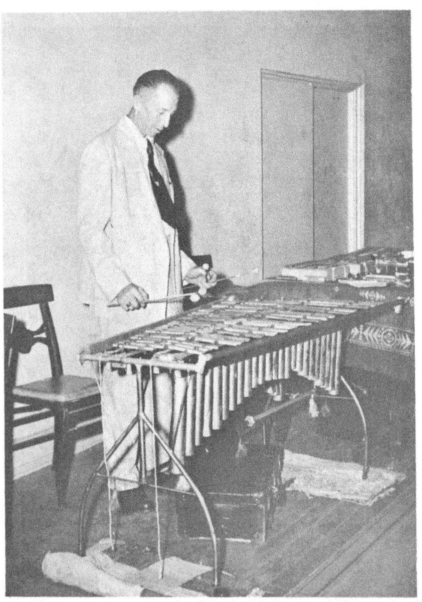

Theodore H. Kleeman and his petrifone made with Colorado petrified wood.

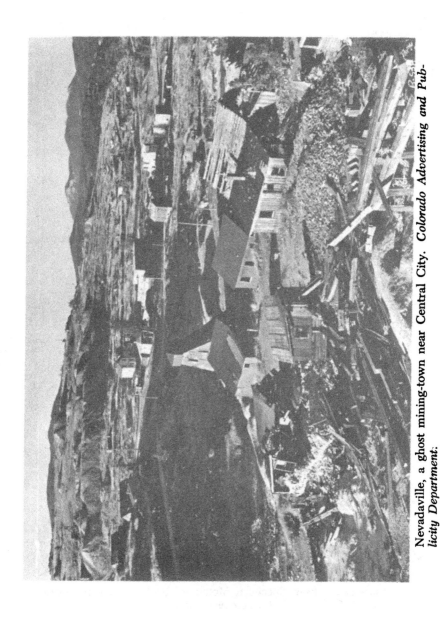

Nevadaville, a ghost mining-town near Central City. *Colorado Advertising and Publicity Department:*

Twinned crystals of aragonite from the eastern foothills of Colorado. *E. B. Eckel, U. S. Geological Survey.*

Agate geodes from Specimen Mountain, Rocky Mountain National Park. *E. B. Eckel, U. S. Geological Survey.*

feet above East Gulch. Road to left goes past mine and above it; do not attempt it.

The deposit was discovered by J. D. Endicott in the early 1900's, and was first reported by Douglas B. Sterrett (Ref. 1-2). Known as the Wild Rose claim, it was worked by C. A. Beghtol for both rose quartz and muscovite mica. The property was then owned and operated by Mrs. Earl E. Zingheim until her death. Two main opencuts and underground workings from them have been employed. A 960-foot inclined tramway was a conspicuous feature of the workings until destroyed for scrap.

The deposit is a pegmatite introduced into gneiss, schist, and other metamorphic rocks. It has two irregularly curving branches and is exposed over an area of nearly 100,000 square feet. The workings are in the eastern, or main, body. Many smaller pegmatites – which are in reality dikes of coarse-grained granite, because they are barren of unusual minerals – occur nearby, together with other small masses of granite. The geology and mineralogy of the deposit have been described by Douglas B. Sterrett (Ref. 3), Kenneth K. Landes (Ref. 4), and John B. Hanley, E. William Heinrich, and Lincoln R. Page (Ref. 5).

The inner zone of the pegmatite consists of separate masses of quartz and microcline feldspar. Surrounding these is a zone of quartz, albite feldspar, and muscovite mica, all intergrown. On the outside is a zone that contains the minerals of both inner zones.

A good deal of the quartz, especially in the center of the deposit, is rose quartz of an attractive color, which varies in depth. Pieces measuring 2 feet or more across are available. The more deeply colored material is found in the center.

Four areas of beryl, some of it being gem aquamarine, have been described from here. This gem material was mentioned a long while ago and seems now to be gone. The well-formed crystals of beryl range in color from bluish white to deep brown, although most are typically greenish blue to pale blue. In size, they average 4 inches across, while some have been as much as 2 feet in diameter and 8 feet in length. The larger ones are embedded in muscovite mica, the smaller ones in albite feldspar. Many of the smaller ones taper toward both ends. A few crystals contain centers composed of an intergrowth of the common pegmatite minerals.

The crystals of columbite-tantalite are small, but one was reported to be 6 inches long. The crystals occur mainly in albite feldspar near the outer edges of the deposit. Microcline feldspar, the commonest mineral, occurs here in enormous crystals, pale pink to light buff in color. A single crystal 75 feet long and 40 feet wide was observed in one exposure. The albite feldspar is mostly light pink. Large masses of grayish-green muscovite mica, as much as 5 feet in diameter, are present, including single books up to 3 feet long and 1 foot wide. Other minerals found in this deposit are garnet, apatite, black tourmaline, biotite mica, magnetite, native gold, and a bismuth mineral. Some of these average about 2 inches in size.

REFERENCES

1. Mineral Resources, 1907, pt. 2, 1908, p. 814.
2. Mineral Resources, 1908, pt. 2, 1909, p. 837.
3. U. S. Geological Survey Bulletin 740, 1923, p. 56.
4. American Mineralogist, vol. 20, 1035, p. 330-331.
5. U. S. Geological Survey Professional Paper 227, 1950, p. 55-61.
6. American Mineralogist, vol. 50, 1965, p. 96-98.

MAPS

Topographic maps: Cotopaxi (1959).
National forests: Rio Grande, San Isabel.

Royal Gorge Park

An area of pegmatites having rather similar compositions lies immediately on both sides of the Royal Gorge, which is the most spectacular part of the 34-mile-long Grand Canyon of the Arkansas River. Sometimes known as Eight Mile Park north of the river and as Webster Park south of the river, this is an irregularly rolling open area of igneous and metamorphic rocks surrounded, except on the north, by sedimentary rock.

The geology and mineralogy of this district were described in detail by E. William Heinrich in two articles (Ref. 1-2), which amplified a more condensed report on the same area by John B. Hanley, Heinrich, and Lincoln R. Page (Ref. 3), and included information on certain minerals given in greater detail in previous articles by Heinrich and C. Wroe Wolfe (Ref. 4-6).

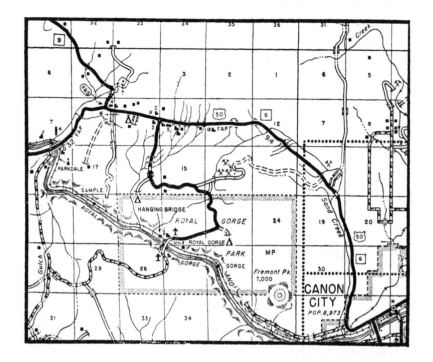

Earlier discoveries and descriptions of the minerals, as distinguished from the general or special geology, were made by W. P. Headden (Ref. 7), who first described the columbite found by E. E. Smith; by W. T. Schaller (Ref. 8-10), who discovered fremontite, which he first named natramblygonite and which is now known to be natromontebrasite (Ref. 11); by Douglas B. Sterrett, who reported new finds in the annual volumes of *Mineral Resources* (Ref. 12-14); and by Kenneth K. Landes (Ref. 15-16).

Dr. Heinrich has classified the pegmatite minerals found here according to the following occurrences:

1. Common rock-forming minerals: Quartz, feldspar, mica
2. Rarer minerals
3. Weathering products

His complete list of minerals, corrected to date, is as follows: Quartz, including rose quartz; microcline, andesine, oligoclase, and albite feldspar, including the cleavelandite variety; muscovite, biotite, and lepidolite mica; chlorite, garnet (spessartite, uvarovite), beryl, tourmaline, fluorapatite, triplite, montebrasite, natromontebrasite, monazite, euxenite, columbite-tantalite, magnetite, hema-

147

tite, chalcocite, native silver, covellite, malachite, azurite, chrysocolla, beyerite, bismutite, limonite, calcite, kaolinite, chalcedony. In addition, cerussite pseudomorphs after natromontebrasite, torbernite or another uranium mineral, and native bismuth are uncertain, and an unidentified manganese oxide is present.

In the rest of this section will be described the best collecting localities of the 18 or more individual quarries and prospects that have been recorded in Royal Gorge Park. Only minerals and rocks of specimen value will be emphasized, each in the locations where they are best developed. Careful collecting offers the possibility of finding new minerals or new places for the same minerals.

The localities that are described are situated north of the Arkansas River, and are logged in order (on both sides of the road) along the Royal Gorge Road from its intersection with U. S. 50. It is suggested that, to follow the logs, you return to the Royal Gorge Road at the same place you left it to reach the preceding locality.

School Section Mine

The log is as follows:

0.0 Intersection of U. S. 50 and Royal Gorge Road.
1.3 Turn right toward Buckskin Joe and park there or at scenic railroad. Quarry is up slope to right between them.

Black tourmaline, beryl, columbite, beyerite, and triplite are among the minerals of especial interest at this mine. There are 2 large open cuts and 10 smaller diggings.

The deposit is an irregular, sheet-like, central pegmatite with several branching offshoots. The main part consists of quartz and muscovite mica grown together, and quartz and microcline feldspar in similar intergrowths, including the curiously attractive pattern known as graphic granite, which resembles the angular writing of ancient civilizations. Surrounding the area of these other intergrowths is a finer grained association of quartz and microcline, with or without muscovite; it contains inclusions of graphic granite. Muscovite and biotite mica have also grown closely together in large blades reaching a maximum length of 6 feet. Scattered within the pegmatite are pods up to 200 feet in length, composed of massive quartz and blocky crystals of microcline up to 6 feet in length.

148

Around the pods are mixed masses of quartz and plagioclase feldspar, in which are found most of the minerals of specimen interest. These include beryl, garnet, tourmaline, apatite, columbite, triplite, beyerite, bismutite, and chalcocite. The tourmaline exists in large clusters of radiating, black crystals, as shown by specimens accumulating on the various dumps. Some of the "tourmaline suns" are as much as 4 feet in diameter, and individual crystals reach 8 inches in length and 3 inches in width. Attractive rosettes of the platy, cleavelandite variety of albite feldspar, along with abundant tourmaline and apatite, are common near the entrance to the farther of the two largest excavations. The apatite occurs in crystals that vary in color from gray green to an iridescent, dark purple and are as long as 8 inches. Triplite in dark-brown, rounded masses as much as 6 inches long, coated black, has been common near the entrance to both of the large cuts and can still be picked up here. Various local segregations of biotite have also been seen, and calcite is concentrated as earthy, cream-colored crusts at the top of the hill.

Van Buskirk Mine

Continuing log:
1.3 Royal Gorge Road, entrance to School Section mine. Continue through fence gate and go to quarry. Or turn around in road entrance just ahead and return.
2.2 Turn left.

Half a dozen pits, opencuts, and trenches represent the main workings of the mine, which borders on a fairly deep gulch.

Red microcline feldspar, white quartz, and biotite and muscovite mica are the chief minerals. Blades of biotite reach a length of 4 feet. A good deal of black tourmaline is concentrated in several places. Some garnet and blue apatite have been collected. Graphic granite of attractive patterns, both coarse and fine, is especially abundant.

Mica Lode

Continuing log:
2.2 Royal Gorge Road, entrance to Van Buskirk mine. Continue toward Royal Gorge.

2.3 Turn left onto mine road.

2.7 Keep left at intersection.

2.8 Gateway to Mica Lode and Meyers Quarry. Closed to vehicles after working hours; danger from mine trucks while operating.

Formerly the largest producer of feldspar in Colorado, the Mica Lode is also a source of abundant muscovite mica, beryl, triplite, pink and black tourmaline, rose quartz, andradite garnet, beyerite, and other desirable minerals. It was once the largest producer of gem tourmaline in the United States outside of California. The mammoth size of the quarry and the rapidly changing nature of the minerals exposed make this a leading attraction among the mineral localities of Colorado.

The exposure and dump on the left nearest the entrance are at the site of the original, modest workings. These have since been extended around the entire hill, which now forms a distinct conical knob with an abrupt south face readily visible from the Royal Gorge Road. In 1906, shortly after the deposit was first opened, this early locality yielded tourmaline of gem quality in pink, lilac, green, and blue colors, as well as colorless and variegated specimens. This material is gone, but occasional light-green, yellow, and pink, gem tourmaline has been found in recent years.

The extensive operation of this mine is indicated by the fact that, together with the adjacent Meyers Quarry described below, it chiefly supported for several years the second largest feldspar mill in the country, which was owned by the Consolidated Feldspar Corporation and is situated at Parkdale, 3 miles west of the intersection of U. S. 50 and the Royal Gorge Road, the point from which the logs in this section of the book begin. The present operator is Lockhart and Sons, of Canon City.

The quarry is constantly changing appearance with the rapid extraction of rock, and so new areas of interesting minerals are being brought to light at unexpected intervals. Specimens are exceedingly abundant on the various dumps.

The Mica Lode is an enormous pegmatite in quartz-mica schist and hornblende gabbro. It is lens shaped and about 1,300 feet long and up to 450 feet wide. On the right (east), it is almost continuous with the Meyers Quarry, which represents the second of the huge cuts that lead off from the main mine road.

A large, central (or core) zone consists of huge masses of white quartz, crystals of pink to red microcline feldspar up to 6 feet in length, and a mixture of plagioclase feldspar and muscovite mica, in which are found beryl and triplite. Beryl is fairly abundant here, and crystals as long as 6 feet have been recovered; their color is usually pale blue green, sometimes yellow green, and their hexagonal shape when intact is often tapered, although many are badly shattered and tend to fall apart. The most numerous beryl crystals have been found toward the left (west) end of the deposit. Pink and black tourmaline are found more or less frequently. The first measurable crystals of triplite ever found anywhere were taken from the Mica Lode, as well as masses as much as 2 feet long. The best place to collect triplite is on the top of the hill along the middle to right (east) edge of the mine. The rare bismuth carbonate mineral beyerite is associated with bismutite in gray nodules in muscovite. Large segregations of greenish muscovite are a prominent feature; they include wedge-shaped blades 1 to 3 feet long, arranged in rosette and comb shapes, and nearly spherical "ball mica" as large as 30 feet across. Considerable spessartite garnet of light-brown to clove-brown color is localized in this part of the pegmatite, where it is intergrown with large masses of chalcocite, which alters to malachite, azurite, and chrysocolla. The first discovery of uvarovite garnet in Colorado was made here several years ago by George Robertson, a student at Fountain Valley School. A large area of very attractive rose quartz, a mineral doubtless previously seen but not earlier described here, was first found in 1953 by the author along the right wall of the quarry, where it widens out.

Surrounding the core is the largest rock unit in the Mica Lode. This consists of intergrowths of quartz and microcline (often as graphic granite) and of quartz and muscovite.

R. H. Magnuson Mine

High on the side of a hill, to the right (south) of the Mica Lode as seen from the road leading to it, is the R. H. Magnuson mine. It can also be reached from the Priest Canyon road, which borders it on the south, although it is not visible from there.

The country rock is a fine-grained, red granite, which has been penetrated by the pegmatite. Graphic granite is the most common

151

material, and there are large masses of pink feldspar and quartz. The specimen material is chiefly yellow beryl, fragments of which are abundant in the float below the pegmatite. Crystals of greenish beryl, some conspicuously large, also occur here.

Meyers Quarry

Directly east of the Mica Lode and now hardly distinguishable from it is the Meyers Quarry. It is likewise operated by the firm of Lockhart and Sons, in the same great combination of workings, it being the second of the big quarries opening onto the main road.

The pegmatite is about ½ mile long, very irregular in shape and enclosed in mica schist and hornblende gabbro. The largest part of the body is very similar to that of the Mica Lode. Within it are a number of isolated masses of quartz and microcline feldspar. "Ball mica," plagioclase feldspar, and crystals of beryl are numerous here.

A substantial lithium content makes parts of the left (west) end of the deposit especially attractive. Patches of coarse cleavelandite (the platy variety of albite feldspar) contain lepidolite (the pink mica) and pink tourmaline, also colored by lithium. At present, the best place for these minerals is high up on the right-hand wall of the quarry, about halfway in. Beryl, columbite, garnet, and natromontebrasite are other interesting minerals occurring here in relatively substantial amounts. The feldspar, the plates of which are as much as 6 inches long, is buff to pink. The lepidolite is found both as pale-lilac plates in large books and as deep-purple flakes, sometimes rimming books of ordinary muscovite mica. The beryl is pale yellow and has the texture of porcelain; some crystals are buried in fine-grained lepidolite, and a few are up to 2 feet in diameter. Nearly white beryl, leached and shattered, may be found in a number of places. The tourmaline is especially attractive, being red, pink, dark green, yellow, blue black, and black. The choicest specimens are watermelon tourmaline – raspberry red inside and green on the outside – set in a mass of varicolored rock. Most of the tourmaline is black; crystals edged with quartz are very common in certain parts of the quarry, especially on the right-hand wall about halfway in. Masses of columbite as heavy as 600 pounds were found here in earlier years. Brown garnet crystals are not uncommon.

The rapidly changing aspect of this mine, as well as the adjacent Mica Lode, makes it futile to state exact places where given minerals occur. The enormous sizes of the pegmatites, however, suggest that interesting material will be readily available somewhere in these deposits for a long time to come. Specimens loosened by blasting are abundant on the dumps.

Border Feldspar No. 2 Mine

Continuing log:

2.3 Royal Gorge Road, entrance to Mica Lode – R. H. Magnuson mine – Meyers Quarry. Continue toward Royal Gorge.

2.5 Turn right where safe immediately after road bank, staying left of trees. Go across upward-sloping field, keeping just to left of fence and avoiding gullies.

2.6 Park beside dump of dark rock beneath Border Feldspar No. 2, which is on ridge above.

This mine is noted for its numerous crystals of black tourmaline. Quartz and graphic granite are very common, and muscovite mica is also available.

At the other end of the same ridge is the Border Feldspar No. 1 mine, which contains quartz, feldspar, muscovite, and graphic granite but no minerals of special interest. Both of these deposits are on land that is part of the Royal Gorge Park, owned by Canon City.

REFERENCES

1. American Mineralogist, vol. 33, 1948, p. 92, 198, 420-448.
2. American Mineralogist, vol. 33, 1948, p. 550-587.
3. U. S. Geological Survey Professional Paper 227, 1950, p. 34-41.
4. American Mineralogist, vol. 31, 1946, p. 198.
5. American Mineralogist, vol. 32, 1947, p. 518-526, 660-669.
6. American Mineralogist, vol. 33, 1948, p. 92.
7. Colorado Scientific Society Proceedings, vol. 8, 1905-1907, p. 57-58.
8. American Journal of Science, ser. 4, vol. 31 (181), 1911, p. 48-50.
9. U. S. Geological Survey Bulletin 509, 1912, p. 101-103.

10. U. S. Geological Survey Bulletin 610, 1916, p. 141-142.
11. American Mineralogist, vol. 40, 1955, p. 1141-1145.
12. Mineral Resources, 1907, pt. 2, 1908, p. 826.
13. Mineral Resources, 1908, pt. 2, 1909, p. 844.
14. U. S. Geological Survey Bulletin 740, 1923, p. 55-56.
15. American Mineralogist, vol. 20, 1935, p. 328-329.
16. American Mineralogist, vol. 24, 1939, p. 188.
17. American Mineralogist, vol. 36, 1951, p.269.
18. Colorado School of Mines Quarterly, vol. 44, no. 2, 1949.
19. U. S. Geological Survey Bulletin 1114, 1961, p. 72, 74, 75, 96, 99, 104, 112, 148, 160, 231, 232-233, 236, 240, 241, 377, 339-340.

MAPS

Topographic maps: Royal Gorge (1959).
National forests: San Isabel.

Cowan Quarries

Deposits near Canon City are still capable of supplying large amounts of specimen travertine. They have been furnishing handsome blocks for numerous structures (especially Federal buildings) throughout the country, and the deserved reputation of this colorful structural stone is spreading. These quarries are readily visible from U. S. 50 on the way toward the Royal Gorge.

The log to this locality from Canon City is as follows:

0.0 Municipal Museum, Sixth and River Streets. Go west on River Street, continuing on U. S. 50 toward Salida.

3.6 Pass one-way Skyline Drive on right.

4.0 Turn right onto mine road and keep left through gate. Gate is locked after 2:30 P.M. and weekends and holidays. If you are locked in, turn around and follow back roads to the right.

4.7 Pass road on left to marble quarry described at end of this article.

6.3 Park in open area at right.

The blocks and irregular pieces strewn around the parking space are mostly waste that was found undesirable because of cracks, excessively large holes, or other defects. Such discarded

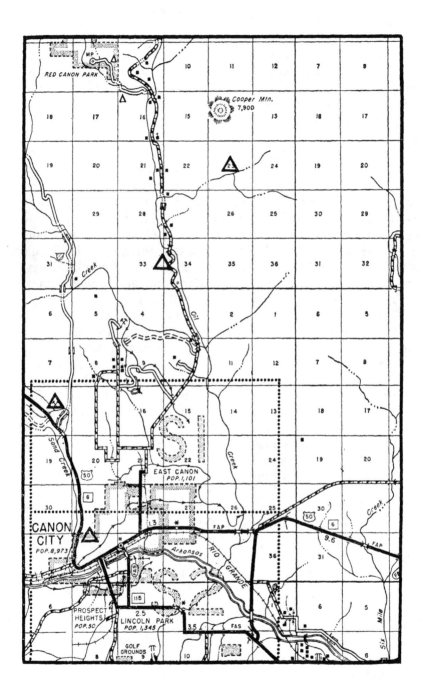

155

material often makes better specimens than the commercial stone, because the porous nature of the travertine, which is its most typical feature, is more evident. Travertine is considered a hot-or cold-spring deposit of calcium carbonate. The cavities may be due to the former presence of vegetable matter buried in the mineralized water. Casts of pine needles and cones are seen in the rock.

The nearby quarry, up the road to the left, is operated by Cowan Brothers (J. E. and David Cowan), who have always been very courteous to visitors who make themselves known. The larger quarry, beyond and to the right, was formerly operated by Colonna and Company of Colorado, Inc. (a subsidiary of Colonna and Company, New York); a long, continuous wire saw, fed with sand as an abrasive, was used to cut the blocks at this quarry. Individual blocks up to 21 tons in weight are extracted and trucked away. There being no processing plant in Colorado, the stone is shipped to Carthage, Missouri, for cutting and polishing.

The Colorosa grade is veined throughout in red, which varies in shade from place to place. This rich, red stone is best seen in the finished slabs employed as facing in such impressive installations as the *Denver Post* Building in Denver, the Shamrock Hotel in Houston, and the New England Telephone Company's building in Boston. The Pentagon Building, in Washington, has some fine pieces of this stone. The Colocreme grade is light brown to cream, with some rose color. The Mahogany grade tends to a rusty color. A gray stone is also mined. Chips for terrazzo flooring are also obtained from this deposit.

Deposits of beautifully colored marble, owned by Cowan Brothers, are situated in this area, one close to the travertine quarries, another several miles up the nearby canyon. The handsome new Fremont County Court House, in Canon City, has splendid installations of this Royal Breccia marble, polished inside the building and rough outside. The newly enlarged Capitol Building, in Washington, has pillars of the same marble. The side road mentioned in the log above leads to the nearer of the marble quarries from which this rock came.

REFERENCES

1. Colorado School of Mines Quarterly, vol. 44, no. 2, 1949, p. 459-464.

Topographic maps: Royal Gorge (1959).
National forests: San Isabel.

Skyline Drive

Selenite crystals have been obtained for many years from the upturned sedimentary rocks that stand east of, and below, the Skyline Drive at Canon City. These strata are locally referred to as pigbacks, in contrast to the great hogback of the Dakota formation, which forms the Skyline Drive. This locality is on public land just outside the Colorado State Penitentiary, and you are required to telephone the prison and tell them that you wish to dig for specimens.

The specimens are found in loose debris at the base of the two hogback loops just north of the last guard tower. In addition to the crystals and clusters of selenite gypsum, there are cephalopod and *Inoceramus* fossils and occasional shark teeth. Cone-in-cone structures are found also. A ride on the one-way Skyline Drive is recommended to give a broad view of the area from above.

MAPS

Topographic maps: Canon City (1959).

Garden Park

Jasperized dinosaur bone and rich agate occur close together in a former dinosaur-bone quarry at the south end of Garden Park, 7 miles north of Canon City. This general area is famous for its yield of skeletons of some of the largest dinosaurs ever to roam the earth. The Canon City Geology Club has been effective in the erection by the State Historical Society of Colorado of a bronze tablet, called the Garden Park Dinosaur Monument, mounted on a travertine base from the nearby quarry and depicting some of the reconstructions from here that are exhibited in noted museums. The description of these fossil beds is given in *Exploring Rocks, Minerals, Fossils in Colorado*, by Richard M. Pearl (Sage Books, The Swallow Press, Inc., Chicago, revised edition, 1969).

The log to this locality from Canon City is as follows:

0.0 Municipal Museum, Sixth and River Streets. Go east on

River Street.

0.7 Turn left on 15th Street toward Red Canons Park.

1.4 Pass Odd Fellows Home on right.

2.1 Turn right onto South Street.

2.2 Turn left onto Park Street toward Red Canons Park.

2.7 Turn right onto High Street.

2.8 Turn left onto Phelps Avenue.

3.1 Keep straight ahead.

4.4 City dump.

Park off road at right. Walk across road and over low knoll, continuing 100 yards or so.

Beyond here, the low-dipping black shale of the Pierre formation contains seams of selenite gypsum. The larger, more solid pieces are found by digging with a pick or shovel 1 foot or so into the soft shale. Some of the crystals coat fossils of Cretaceous cephalopods. Crystals of white calcite and brown barite several inches long occur in septarian concretions here. Cone-in-cone structure is prominent. Robert L. Chadbourne, of Colorado Springs, first showed this locality to the author.

Continuing log:

6.2 Pass site of second oil-well locality in United States, across Oil Creek. Black, oil-saturated layer at base of cliff at water's edge. Bronze marker in red marble on left of road.

8.0 Pass site of 1954-1957 dinosaur excavations by Cleveland Museum of Natural History on right, across stream, at base of cliff.

8.2 Garden Park Dinosaur Monument on left. Park here, walk up road across culvert at mouth of gully, and climb steep bank on left side of road, following old haulage road.

The bed of sedimentary rock that contains the bones is a few hundred feet above the one that contains the agates, and both are in the Morrison formation of Jurassic age. Uranium deposits were worked briefly in the same formation elsewhere in Garden Park.

The silicified bone is red, brown, and yellow, with gray and white matrix. Some of it takes a good polish and is attractive enough for gem use. The rest, however, has been replaced by minerals of varying hardness and is difficult to polish evenly.

The agate is in the form of geodes, seams, and irregular pieces. It ranges from white to brown, and includes as well gray, yellow, orange, and bright-red colors; some specimens are varicolored. The St. Stephen variety — translucent with round, red spots — is particularly fine. The patterns of the agate banding, combined with rich colors that need no artificial dye, make this deposit an unusually interesting one. Petrified shells and fibrous, pink calcite occur here also.

Douglas B. Sterrett (Ref. 1) in 1909 was the first to mention the locality, which had been worked to a small extent by J. D. Endicott, and he later described it more fully (Ref. 2). No collector has done so much here as Frank C. Kessler, executive secretary of the Canon City Geology Club, who was responsible for important recoveries of dinosaur remains from the heart of the Garden Park area.

REFERENCES

1. Mineral Resources, 1908, pt. 2, 1909, p. 807.
2. Mineral Resources, 1910, pt. 2, 1911, p. 849-850.
3. U. S. Geological Survey Folio 7, 1894.

MAPS

Topographic maps: Cooper Mountain (1954).
National forests: San Isabel.

Felch Creek

In the badlands adjacent to Felch Creek, beyond the Garden Park locality described above, occur numerous geodes, the choicest containing exquisite crystals of quartz, calcite, barite, celestite, and goethite. Many specimens of red jasper, some red agate, large nodules of alabaster, other varieties of gypsum, and quantities of petrified dinosaur bone are also found here.

The log to this locality continues from the one to Garden Park, as follows:

8.2 Parking space at Garden Park Dinosaur Monument. Continue north.
8.9 Pass road on right to square, block farmhouse, and turn right onto faint road through fence gate. Close gate.
9.2 Go through fence gate and close gate. Stay right on rough road at next fork.

159

9.8 Park in open area at end of road. Walk into dry gullies ahead.

Abundant material is usually available as a result of weathering and the activities of collectors, but digging is sometimes required and is recommended because of the thick overburden.

This excellent locality was discovered during a high school field trip about 1940 by Frank C. Kessler and was described and mapped by Glenn R. Scott (Ref. 1).

Although the amount and variety of specimens are most satisfying, the geodes are of outstanding interest. They occur in a siliceous layer, 2 to 6 inches thick, within the colored shales of the Morrison formation, of Jurassic age; when the hard rock weathers, the geodes are released and roll downhill. Inside the geodes are small, sparkling crystals of colorless, yellow, or pale-amethyst quartz. Most of the geodes also contain pointed crystals of yellow calcite. Crystals and cleavable pieces of barite, which is transparent and amber colored, and celestite, which is colorless, milky, pinkish, or blue, are found in some of the geodes. The most attractive specimens are delicately sprinkled with brown needles of goethite, which penetrates the other minerals. Millerite may also have been found here.

The lumps of pink alabaster, together with satin spar, selenite, and sugary gypsum, are similar to those occurring in the same sedimentary formations elsewhere along the eastern foothills of the Front Range. The red jasper and agate are of good cutting quality.

REFERENCES

1. Rocks and Minerals, vol. 24, 1949, p. 142-143.
2. U. S. Geological Survey Folio 7, 1894.
3. U. S. Geological Survey Bulletin 1114, 1961, p. 20, 93, 166, 225, 277.

MAPS

Topographic maps: Cooper Mountain (1954).
National forests: San Isabel.

Curio Hill

Among the choicest of Colorado agates, in their attractive

natural colors, translucency, and delicate patterns, are those that have been found on the top and sides of Curio Hill, a low limestone ridge 7 miles south of Canon City. Also called Specimen Ridge, Curio Hill is in the southern part of Fremont County, at the northern end of the Wet Mountains.

The log to Curio Hill from Canon City is as follows:

0.0 Municipal Museum, Sixth and River Streets. Go west on River Street.

0.2 Turn left onto 4th Street Viaduct and bridge across Denver and Rio Grande Western Railroad and Arkansas River, toward Westcliffe.

0.9 Turn left onto Highland Avenue.

1.1 Keep right across railroad tracks.

1.3 Prospect Heights Mercantile Store. Jog left and then right toward Westcliffe.

1.5 Keep left at road junction.

3.0 Keep right onto gravel at road to Cotter Corporation mill.

5.9 Join Oak Creek road coming from left.

6.4 Pass entrance to ranch on left.

7.1 Stop and find moveable gate nearby to walk through.

This ridge rises 150 feet above the surrounding land, to an altitude of 6,300 feet; it runs about ¼ mile in a northwest direction. A bed of white sandstone outcrops near the highest part of the ridge and abuts against red granite on the west.

Blasting was done some years ago along the top of the Ordovician limestone outcrop that constitutes most of Curio Hill, and many agates have also been taken from the loose gravel and soil on the sides and at the base of the ridge. The best place to collect is between the road and the old pit, both north and south of the pit.

The agate occurs in globular pieces rarely as much as 2 inches in size, and in veins and cavities, together with quartz and calcite crystals. Almost all the agate is banded, some specimens having alternating narrow and wide bands. The colors range from white to brown, and include gray, yellowish red, and bright red; several colors may be seen in the same piece.

This locality was first mentioned by Douglas B. Sterrett (Ref. 1) in 1909 and described by him in more detail in 1911 (Ref. 2), although the place had been known to collectors for many years previously and had been worked by J. D. Endicott. It was first made known to the author by Frank C. Kessler, who guided Canon City school trips here.

REFERENCES

1. Mineral Resources of the U. S., 1908, pt. 2, 1909, p. 807.
2. Mineral Resources of the U. S., 1910, pt. 2, 1911, p. 848.
3. U. S. Geological Survey Bulletin 1114, 1961, p. 19, 277.

MAPS

Topographic maps: Rockvale (1959).
National forests: San Isabel.

Penrose

An easily accessible locality for calcite is situated 11 miles east of Canon City. It was discovered by Cris W. Christensen, of Colorado Springs, and made known to the author by Robert L. Chadbourne, of Colorado Springs. It is on school ground and open.

The log to this locality is as follows:

0.0 Intersection of U. S. 50 and Colo. 115 south of Penrose. Go south on Colo. 115 toward Florence.

1.5 Park on left of road near foundation of old building. Walk back and across road; go through fence and up valley about 150 yards, keeping on the left side. Then go up first side gully 25 feet.

The calcite lies scattered on the ground and can be dug from many seams in the soft limestone. Crystals of several habits, up to 4 inches in length, occur facing inward in cavities in the rock, and others face outward on both sides of the thin-bedded Niobrara limestone. There are scattered veins throughout the general area on both sides.

On the right side of Colo. 115, at several places north of Penrose, toward Colorado Springs, there are several places north of Penrose, toward Colorado Springs, there are several outcrops of gypsum that contain alabaster. These are in the Lykins Formation and are close to the road and easily recognized.

162

Topographic maps: Florence (1959).

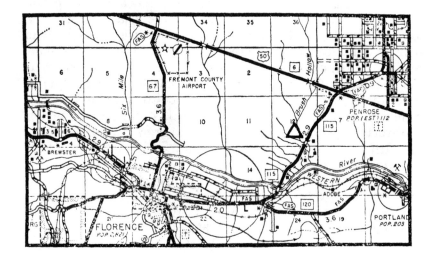

The highway between Trinidad and Lamar goes through Las Animas, Otero, Bent, and Prowers Counties, at the border of Kansas. The Southeastern Plains region of Colorado is characterized in the so-called Raton section by high mesas capped by basalt, sharp ridges, and steeply cut canyons. The Van Bremmer dike locality, near Trinidad, is described below. Much agatized dinosaur bone has been collected on private ranch land south of La Junta, in Otero County. One of the principal commercial sources in Colorado for alabaster for ornamental objects is about 30 miles south of La Junta; the Colorado Alabaster Company, of that city, manufactures ornamental objects from it. A tire-service building constructed by W. G. Brown in Lamar, in Prowers County, is built of petrified wood found 18 miles south of the city, along U. S. 287-385. The Lamar Lumber Company building, also constructed by Mr. Brown, was formerly built of petrified wood, as reported by Pearl Anoe (Ref. 1).

Monument Lake, 35 miles west of Trinidad on Colo. 12, is one of the pretty places in the state. The upturned sedimentary strata at Stonewall make fine scenery. In this area is the green jasper

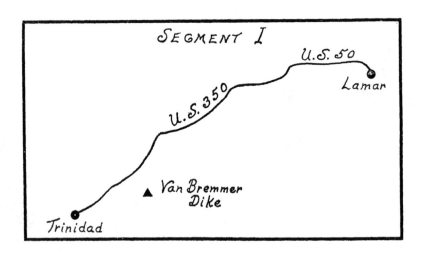

known as "Stonewall jade," which could be looked for in stream gravels and along the road, especially near Monument Lake. This occurrence was mentioned by R. M. Tatum (Ref. 2).

The Culebra Peak Rock Hound Club meets in Trinidad on the third Thursday of the month at Trinidad State Junior College.

REFERENCES

1. Colorado Wonderland, vol. 7, no. 4, 1956, p. 26.
2. Southwestern Lore, vol. 12, 1946, p. 58.

MAPS

The Index to Topographic Maps of Colorado, obtainable free from the U. S. Geological Survey, Federal Center, Denver, Colorado 80225, shows which parts of southeastern Colorado are covered by topographic maps.

Van Bremmer Dike

A large amount of calcite, in the form of crystals and veins, can be found in shale along the sides of the Van Bremmer dike northeast of Trinidad, in Las Animas County. It fills veins in concretions and septaria, being dark brown to green, and it occurs as crystals – some of which are clear enough to be Iceland spar – on the inside of them. Evidently referred to in older literature, this locality was described by R. M. Tatum (Ref. 1), who also mentions the presence here of barite, but the author did not identify any. The shale is the Carlile Formation, belonging to the Benton Group, of Cretaceous age. The formation beneath it is the Greenhorn Limestone, likewise of the Benton Group; at this place, it is thickly fossiliferous with shells and coiled ammonites. These sedimentary rocks are supported by the basalt of the dike, which crosses the road.

This locality can be reached from Model, on U. S. 350, 19 miles northeast of Trinidad, according to the following log:

0.0 Texaco station at Model. Go north on U. S. 350 toward La Junta.

0.7 Turn right on dirt road toward Canyon Station.

4.3 Keep ahead at junctions.

6.7 Cross cattle guard and keep left at fork.

8.0 Turn off road to right just before wood bridge, and drive toward hills, keeping to right of fence. Specimens on middle slopes of knolls whitened by broken shale.

REFERENCE

1. Southwestern Lore, vol. 12, 1946, p. 41-42, 49, 58.

MAPS

Topographic map: Elmoro (1895).

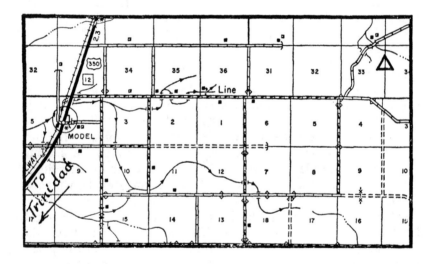

COLORADO SPRINGS AREA

Long the best known place in Colorado, Pikes Peak has given its name to the labels of many fine mineral specimens that have been found up to 40 miles from the mountain itself. In recent years, the individual localities have been more frequently referred to by their own names, such as Crystal Park, Crystal Peak, Devils Head, and others described in this book. The deposits that are associated with the Pikes Peak granite are much alike, however, and a good deal of the mineralogy and geology is similar for the several individual localities that are mentioned. Glen Cove, with its superb crystals of blue topaz, is the only significant locality on Pikes Peak itself (Ref. 1).

In addition to those minerals found in the "hard rock" of the mountains, many specimens of the Pikes Peak region are collected from the sedimentary rocks of the foothills and plains. Along the eastern edge of the Front Range are the uptilted sedimentary rocks known as the Foothills, which merge into the flat rocks of the Great Plains, stretching to the Mississippi River and beyond. Thus, three divisions of topography — mountains, foothills, plains — are represented in what is popularly regarded as the Pikes Peak region, with Colorado Springs as its capital city.

Collectors are invited to attend the meetings of the Colorado Springs Mineralogical Society, held on the second Friday of each month from September to July at the First Christian Church, North Cascade and East Platte Avenues.

Members of the society have contributed the specimens exhibited at the Pioneers' Museum, 25 West Kiowa Street, where there is a separate room of fluorescent minerals. Hours are 10:00 to 5:00 from Monday to Saturday and from 2:00 to 5:00 on Sunday. Probably no other city of comparable size in the world can show private collections of crystals equal to those in Colorado Springs, and the collections of micromounts are also impressive. J. K. Halliburton (Ref. 2) described some of the most important ones of several years ago.

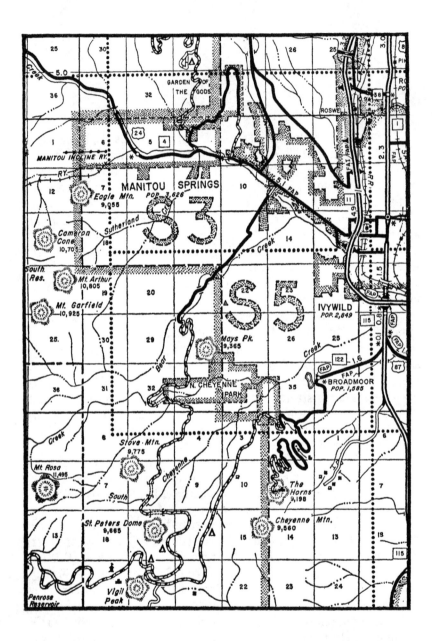

REFERENCES

1. Gems and Minerals, no. 309, 1963, p. 16-19.
2. Rocks and Minerals, vol. 22, 1947, p. 722-723.

Crystal Park

As one of Colorado's oldest and best collecting localities, Crystal Park has really been the source of many of the gem minerals labeled Pikes Peak or Manitou. In turn, it has been credited with specimens found elsewhere in the region. Abundant smoky quartz and rock crystal, some gem topaz and phenakite, and a large part of the amazonstone for which Colorado is famous have been taken from Crystal Park and from an area toward the northwest (on the slopes of Cameron Cone) and toward the southeast, as described by Edwin W. Over, Jr. (Ref. 1). Dozens of prospect holes are in the area, and dumps may be seen in many places. Most of the work has been done by men living in nearby towns.

Crystal Park is a relatively level, steplike area in the part of the Pikes Peak region that lies at the base of Cameron Cone, about 2 miles southwest of Manitou Springs and 6 miles west of Colorado Springs. A wooded flatland, Crystal Park is a sort of terrace, standing above the sloping land that follows the course of Sutherland and East Sutherland Creeks northwestward into Engelmann Canyon, and lying below the steep masses of Cameron Cone (10,709 feet), Mount Arthur (10,807 feet), and Mount Garfield (10,925 feet) on the west and south. These three mountains are connected successively by ridges. Crystal Park extends for about 2 miles in a southeasterly direction, and has an altitude of about 8,600 feet.

Fallen into disrepair a few years ago, so that nothing wider than a horse could follow it in some places, the Crystal Park Auto Road has been rebuilt as a good toll road, open during the summer. It begins on West Colorado Avenue, 2.0 miles west of 26th Street, between Colorado Springs and Manitou Springs, and goes west past McLaughlin's Lodge for about 8 miles; the trip takes about 1 hour. Crystal Park is promoted from time to time as a tourist resort. Construction of facilities for visitors may encroach upon the former collecting areas but should make for greater convenience in opening up new deposits. The motorist may be advised that collecting is not allowed, but most of the land belongs to the U. S. Forest Service, and prospecting on such land is of course permitted.

In addition to this road, trails from a number of directions

169

reach Crystal Park. The most accessible trails begin at McLaughlin's Lodge; between Red Mountain and Iron Mountain, south of Manitou Springs; and from the Boy Scout Camp, past the junction of Hunters Run and the Bear Creek Canyon road.

The gem specimens of Crystal Park occur in cavities in pegmatite in the Pikes Peak granite, which covers the region, and especially in a coarse, white phase of the granite. The cavities vary in width up to 1 yard and are irregularly distributed, some being close together and connected by small veins, although others are quite isolated. The usual pegmatite minerals — quartz, feldspar, and mica — are the most common.

Amazonstone occurs in crystals that range in width up to several inches, being as large as those from other Colorado localities. The best specimens obtained here since the early days have been found in recent years by Henry E. Mathias, of Colorado Springs. Associated with the feldspar are the gemmy crystals of smoky quartz and rock crystal, and (less frequently) topaz and phenakite, while zircon is rare. In 1954, Edwin W. Over, Jr., secured some magnificent crystals in pinkish topaz several inches long from pockets on the southeast side of Cameron Cone. The same winter, a number of the author's students picked up dozens of quartz crystals newly released by weathering of the rocks on the northeast side of the peak.

Many fine specimens of phenakite have been taken from Crystal Park. Many more have probably been overlooked because of their close resemblance to quartz. Most phenakite from Crystal Park is colorless, but some is pale yellow and gray. The crystals are rhombohedral in habit; some faces are smooth and bright, while others are striated and etched. In 1884, 50 good crystals were obtained, more than a quarter of which were of gem quality, the largest measuring 3 inches across. What was described by Whitman Cross and W. F. Hillebrand (Ref. 2) as the finest phenakite discovered in the United States was mined at Crystal Park in 1887 and placed in the collection of Whitman Cross; it weighed about 460 carats and had transparent spots in it.

Other minerals in this area include fluorite, columbite, biotite, hematite (in quartz and as pseudomorphs after siderite), hornblende, and black tourmaline, the latter two also enclosed in quartz.

REFERENCES

1. Rocks and Minerals, vol. 4, 1929, p. 106-107.
2. American Journal of Science, ser. 3, vol. 24 (124), 1882, p. 282.
3. Mineral Resources, 1883-1884 (1885), p. 724.
4. Mineralogist, vol. 9, 1941, p. 123-124.
5. American Journal of Science, ser. 3, vol. 33 (133), 1887, p. 131.
6. U. S. Geological Survey Bulletin 20, 1885, p. 68-69, 70-71.
7. U. S. Geological Survey Folio 203, 1916.
8. American Journal of Science, ser. 3, vol. 41, 1891, p. 439.
9. U. S. Geological Survey Bulletin 1114, 1961, p. 18-19, 66-67, 140-141, 146, 253, 272, 334.
10. Gems and Minerals, no. 309, 1963, p. 16-19.

MAPS

Topographic maps: Colorado Springs (two scales, 1948-61), Manitou Springs (1948-61), Pikes Peak and Vicinity (1956, contour or shaded relief edition, $1.00).
National forests: Pike.

St. Peters Dome District

The conical summit of St. Peters Dome is thrust conspicuously into the sky to an altitude of 9,665 feet along the Gold Camp Road, between Colorado Springs and Cripple Creek. Situated beneath the towering majesty of Pikes Peak, this is one of the outstanding mineral localities of the world, noted for its many rare species and the good quality of some of its more common ones. Stove, or Cook Stove, Mountain, attaining the even higher altitude of 9,782 feet, is generally regarded as part of the same mass, though separated from the better known St. Peters Dome itself by South Cheyenne Creek. Across Rock Creek, to the south, is Sugarloaf Mountain (about 9,600 feet), and then, separated from it by Little Fountain Creek, is Mount Vigil (10,075 feet). These peaks of the so-called St. Peters Dome district are described here together, because they are so similar and are reached by the same route, which nearly encircles St. Peters Dome. Most of the localities described are under the jurisdiction of the United States Forest Service and are on public land, though open to the staking of mining claims; the Eureka tunnel and Duffield mine, and an occasional other property, are privately owned, though not at present (1971) occupied. The area of St. Peters Dome has been expanded toward Mount Rosa as a result of the prospecting for radioactive minerals between 1950 and 1958; the chief concentrations are 1 mile north of Rosemont and 2 miles east of Mount Rosa. The latest descriptions of the minerals of St. Peters Dome is by Eugene B. Gross and E. William Heinrich (Ref. 28).

The pegmatites are bodies of coarse-grained rock enclosed within the Pikes Peak granite. The essential minerals are typical, pink microcline-perthite feldspar and ordinary quartz, biotite mica, oligoclase feldspar and hornblende. Accessory minerals include apatite, magnetite, zircon, sphene, and allanite; if fluorite is an accessory mineral rather than a later introduced one, it is the most interesting of these. The individual pegmatites are very numerous, with a great many more of them undoubtedly awaiting recovery. The most experienced collector in the area, Edwin W. Over, Jr., had himself examined nearly a thousand pegmatites on St. Peters Dome. Except when lying on the surface or embedded in the soil, specimens can be secured only by penetrating the rock with the aid of tools.

Gem-quality zircon, amazonstone, smoky quartz, rock crystal, phenakite, and topaz are found in the pegmatite, both frozen to the wall and open (even loose) in the pockets, or so-called miarolitic cavities. All these minerals except zircon have a similar occurrence throughout the Pikes Peak region, being better in some places, poorer in others. Gem zircon, however, is known (except in microscopic sizes) only on St. Peters Dome. In the famous Eureka tunnel locality, small but beautifully colored and superbly formed crystals, of purple and other colors, as fine as any of the early-day ones, have been picked up by members of the author's parties. Some of the best specimens have been found here by Mr. Over, George M. White, Willard W. Wulff, J. Perry Osborn, Orville A. Reese, and Timothy C. Anglund, all local collectors.

The fluorine-bearing minerals are exceptionally interesting. An abundance of fluorine in the original molten rock that solidified to become the Pikes Peak granite, or in the later solutions that coursed through it, is freely indicated by the purple fluorite in the country rock and in the veins and pegmatites that traverse it. Small pieces lie everywhere in the gravelly soil. The water of the

Colorado Springs area is so heavily charged with fluorine that it frequently produces mottling of enamel on teeth, at the same time reducing decay. Fluorite for the steel industry has been mined from a number of veins scattered throughout the slopes of St. Peters Dome, many of which are easily visible from the road, and some of which are mentioned below. The cleavable, purple masses of fluorite make attractive specimens. A little green and white fluorite is present. Accompanying fluorite in the veins are galena, sphalerite, pyrite, chalcopyrite, pink barite, specular hematite, quartz (some chloritic), and chalcedony. Gold and silver have been won from the quartz-fluorite veins, probably coming from the sulfide minerals mentioned, as explained by Thomas A. Steven (Ref. 1).

Weberite is the next most important fluorine mineral, as reported by Hans Pauly (Ref. 2), constituting most of what had been thought to be cryolite. Visitors from the huge cryolite quarry in Greenland and the processing plant in Denmark have come to St. Peters Dome to see the feeble remnants of what was said to be the second largest source (but not a commercial one) of cryolite.

Most of the other fluorine minerals are very rare, hard to find, and equally difficult to recognize. They occur mostly with weberite or cryolite, some being direct alteration products of them. These have curious names, such as pachnolite, gearksutite, prosopite, bastnasite, thomsenolite, ralstonite, elpasolite, and fluocerite (here called tysonite). The last two names have a local significance, elpasolite being named for El Paso County, and tysonite for its discoverer, S. T. Tyson.

Additional minerals in the pegmatite include astrophyllite, riebeckite, pyrochlore, cassiterite, rutile, fergusonite, xenotime, fayalite, chlorite, genthelvite (formerly referred to here as dana-lite), columbite (perhaps the first discovery in Colorado), galena, sphalerite, chalcopyrite, hematite, pyrite, molybdenite, aegirine, thorite, microlite, monazite, kasolite, lanthanite, bertrandite, sericite, and perhaps doverite. Some of these, such as astrophyllite (described below) are exceeding rare elsewhere in the world but abundant here; this was the first locality for astrophyllite in the United States. Others, such as cassiterite, are rare here though common enough elsewhere. The first fergusonite in Colorado was found at St. Peters Dome by J. D. Custer, of Colorado Springs, who was also the finder of prosopite and pyrochlore in the

173

cryolite veins. This district is the only known occurrence of genthelvite in the world (Ref. 3); the largest crystal was found in 1949 by John W. Adams (Ref. 4); Glenn R. Scott (Ref. 23) has described the crystallography.

Sharp, pyramidal crystals of opaque, brown zircon, of no gem value, are widespread in the veins and pegmatite pockets in the Pikes Peak granite. Zircon is a minor constituent of the granite itself. The crystals occur as individuals and in clusters that look as though they were stacked inside one another. The best ones are found at the contact between quartz and the surrounding feldspar; the cleanest ones are embedded in quartz.

Except the first locality, the following ones in the St. Peters Dome district are logged along Colo. 33 toward Cripple Creek from the metal gate at the intersection of the Gold Camp Road (Colo. 336) and the one-way High Drive. This point is above Helen Hunt Falls, in North Cheyenne Canyon. It is directly accessible either from West 26th Street (by going west on Colorado Avenue or the freeway from Colorado Springs) or from Cheyenne Boulevard (Colo. 226), which diverges to lead up North Cheyenne Canyon to Helen Hunt Falls and then rises to the intersection. The Gold Camp Road, formerly the Corley Mountain Highway, is constructed on the grade of the abandoned Colorado Springs and Cripple Creek District Railroad, familiarly known as the Short Line.

0.0 Intersection of Gold Camp Road and High Drive. (Descending road goes toward Helen Hunt Falls.) Go down.

0.3 On the right of road, across canyon, is an old tunnel.

On the dump, the largest zircon crystals in the Pikes Peak region were found in 1946 by Warren Johannsen. Many hundreds of them, all frozen in the rock, were recovered by George M. White and J. Perry Osborne, both of Colorado Springs. The crystals are opaque, chocolate brown to metallic black, and up to 2 inches in height and 1½ inches in width.

Continuing log:

0.0 Intersection. Go toward Cripple Creek on Colo. 336.

0.6 Park at large turnoff at forest sign at right of road. Climb hillslope above road.

This locality seems to be an erratic one, yielding good material at times but nothing at other times. Hematite pseudomorphs after

174

siderite up to 2 inches across have been found loose here. Rock crystal and smoky-quartz crystals are also known from here.

Continuing log:

0.6 Parking place mentioned above. Continue.

1.2 Go through first tunnel.

2.4 Fairview, former site of railroad station. Park off road in open space at right. View of Colorado Springs between mountains. Summit of St. Peters Dome visible beyond curve ahead. Localities above and below road.

The slopes below the road, bearing toward the right, are believed to be the best locality for bastnasite. This is possibly the original locality for bastnasite and tysonite (fluocerite), reported to have been lost after the death of several of the local collectors (Ref. 5). Some excellent crystals have come from here during the past few years. Bastnasite is a reddish-brown, rare-earth mineral occurring in square crystals, sometimes in parallel growth with tysonite, sometimes alone.

Toward the left of the bastnasite locality is a deposit of fluorite. On the slopes above the road and toward the right are diggings from which quartz crystals have been taken.

Continuing log:

2.4 Parking place at Fairview. Continue.

2.9 Park off road at right (one car only). Stove, or Cook Stove, Mountain locality is reached by climbing low ridge to the right (north). From low point on the ridge, two trails go about ¼ mile to Buffalo Creek Valley. Continue up creek to open gravel slopes on left.

Amazonstone and smoky quartz are the best known minerals here. Bastnasite, fergusonite, fluorcerite (originally tysonite), fluorite, genthelvite, lanthanite, phenakite, topaz, and zircon have also been recorded from this locality by Edwin B. Eckel (Ref. 6.)

Continuing log:

2.9 Parking place mentioned above. Continue.

5.0 Go through second tunnel.

5.3 Pass side road on left.

5.8 Just before third tunnel, at road junction, park on right (off both roads). Walk up road to left as far as first main curve in new Forest Service access road, continue in same

175

direction on trail as far as cement building on left, then down slope to right 100 feet to dump and broken tunnel of old mine.

Here, and to a lesser extent at the Eureka tunnel, may be collected more cryolite than at any other known place on St. Peters Dome, perhaps more than anywhere else outside of Greenland or the Soviet Union, although the amount has fast become extremely limited. Furthermore, most of it has been shown to be weberite. The cryolite is pink and pale green when entirely fresh, but practically all of it resembles solid paraffin or translucent quartz, from which it can be distinguished by scratching it with a knife (quartz is hard). Recognition of the cryolite (or weberite) is aided by its banded appearance, which is due to alteration along planes of parting. Any touch of color disappears upon heating, leaving the cryolite (or weberite) pure white.

Milky quartz, microcline feldspar, riebeckite, and other pegmatite minerals are available on the ground around the old mine. Columbite has been described from here. Small specimens of minerals produced by the alteration of cryolite are recognized occasionally by those who find them. Pachnolite (a white or pale green, opaque mineral) and prosopite (usually a colorless, glassy mineral) are the most prominent of these, and purple or green fluorite is abundant. Later changes have left chalcedony and kaolinite. These minerals were mentioned as early as 1877 by George August Koenig (Ref. 29), and good descriptions were given in 1882 by Whitman Cross and W. F. Hillebrand (Ref. 7-9) and in 1935 by Kenneth K. Landes (Ref. 10).

From the edge of the dump, one can follow an obscure trail between several high-crowned trees to the right, keeping nearly to the same altitude for about 1/3 mile. On the near side of the first stream is the Eureka tunnel. From there, the main highway is reached by a short but very steep climb. This locality for gem zircon can also be reached from the road as directed below.

Continuing log:

5.8 Parking place mentioned above. Continue through tunnel.

6.1 Park off road on either side. Eureka tunnel reached on left side of stream about 400 feet below road.

The superb zircon crystals formerly obtained by collectors from the Eureka tunnel were as noteworthy for their excellent form as for their beautiful colors in pink, emerald green, light wine to honey yellow, purple, and rich reddish brown. They were not large, ranging in size up to about ½ inch, and the best specimens were usually the smallest, but small gemstones could be cut from them. The crystals were chemically pure and remarkably transparent. The zircon occurred in quartz, kaolinite, and yellowish mica; the last two minerals had altered from the microcline feldspar in which the zircon was originally enclosed, and served to protect the gem and make removal easy.

Continuing log:

6.1 Parking place mentioned above. Continue.

6.6 Having gone between second high roadbank, park off right side of road beyond the turn. Cheyenne Mountain (lodge on top) visible as higher ridge directly east across wide canyon. Climb down steep but safe gravel slope below road and take ¼-mile road to left. Or walk down this Forest Service access road (impassable for cars) where it reaches the highway just beyond parking place. Stay right at first fork, then straight ahead when road turns right.

This locality is clearly marked by prospect pits. Until 1954, when mining-claim notices were posted and most of the loose surface material removed, zircon, riebeckite, and astrophyllite were especially abundant here. The riebeckite is dark bluish black and resembles common hornblende, to which it is closely related, being a sodium amphibole. The crystals are sometimes a foot in length and several inches in width. In October 1948, Harold Hofer, Jr., formerly of Denver, who was then 14 years old, found here one of the finest complete specimens of riebeckite ever known. This mineral was originally called arfvedsonite, another sodium amphilbole.

The astrophyllite, also a complex, sodium-bearing silicate, occurs in beautiful, gold or bronze crystals having a bladed habit. Some of it radiates in a starlike pattern, and some of it forms aggregates associated with quartz and zircon. It might easily be mistaken for weathered biotite mica by anyone not expecting to find anything so unusual as astrophyllite. Considerable brown zircon, masses of hematite, and large numbers of crystals of cleavage pieces of microcline are other minerals of interest to collectors. The abundant quartz here has little value.

A broad view of Colorado Springs opens out from this place. The thumb-shaped rock in the middle distance is Specimen Rock (10,093 feet), the base of which has long been noted for good amazonstone. From there came the first topaz in Colorado and the first phenakite in the United States. Next to Specimen Rock on the north is Sentinel Rock, also part of the area referred to locally as Bear Creek or Bear Canyon, and typical of the rest of the Pikes Peak region, as far as mineral specimens are concerned. Edwin W. Over, Jr., found superb topaz crystals there in recent years. Access to the localities mentioned in this paragraph is usually from the High Drive, and permission should be secured at the City Engineer's office in the Colorado Springs City Hall.

Continuing log:
6.5 Parking place mentioned above. Continue.
9.1 Junction of Colo. 336 and Colo. 122 (left-hand road).

Beginning at Broadmoor (southwest of Colorado Springs), to the right of the arch that leads to the top of Cheyenne Mountain, and generally paralleling the Gold Camp Road (Colo. 336) at a lower elevation, is the Cripple Creek Road (Colo. 122, the old Cripple Creek Stage Road.). This winds up the side of Cheyenne Mountain and joins the Gold Camp Road at Sugarloaf Mountain, south of St. Peters Dome. It was the earliest thoroughfare to Cripple Creek. An enormous smoky-quartz crystal was found about 1941 along this road, between Cheyenne Mountain and St. Peters Dome, by Ira Hofer, of Knob Hill. Broken after it was found, it weighed 46 pounds in addition to the lost terminations, and so it is perhaps the third (or fourth) largest quartz crystal ever discovered in Colorado.

Continuing log:
9.1 Junction mentioned above. Continue.
10.0 Duffield lookout at base of St. Peters Dome. Walk back on parallel road, formerly main Gold Camp Road, about 1/3 mile to Duffield fluorspar mine.

Since 1910, this has been a substantial producer of fluorspar from time to time, especially in 1944 and 1945. Specimens of rich purple and green color are abundant in the workings and on the dumps, along with gray and pink barite, white quartz, and a little sphalerite and galena.

The workings on the left (north) of the road are known as the Leyte pit. Those of the right (south) belong to the Hughes Boss shafts, which furnished the first commercial fluorite in the St. Peters Dome district. Together, they extend about 400 feet and are about 20 feet in width. These deposits have been described briefly by Harry A. Aurand (Ref. 10) and Thomas A. Steven (Ref. 1).

Colo. 336 continues on to Cripple Creek, one of the great gold camps of history. Its specimens of calaverite, krennerite, and other gold and silver tellurides, attractively colored with purple fluorite, are prizes for any mineral collector. Turquoise is also found here; look on Mineral Hill, north of the hospital, and on the hill opposite the cemetery and fairgrounds.

Most of the natural-appearing specimens of gold that are offered for sale at Cripple Creek have actually been "roasted" to drive off the tellurium and leave the precious metal on the surface. Even a spectacular-looking specimen, however, may contain only a small amount of gold. The tellurides are chiefly silver colored before being roasted; afterward, they generally show tiny craters where the gas has escaped, something unknown in native gold.

REFERENCES

1. Colorado Scientific Society Proceedings, vol. 15, 1949, p. 259-284.
2. American Mineralogist, vol. 39, 1954, p. 669-674.
3. American Mineralogist, vol. 29, 1944, p. 163-191.
4. American Mineralogist, vol. 38, 1953, p. 858-860.
5. Rocks and Minerals, vol. 4, 1929, p. 106-107.
6. U. S. Geological Survey Bulletin 1114, 1961, p. 19, 48-49, 61, 66-67, 112, 119, 141, 146, 151, 163, 200, 259, 267, 281, 289, 300, 334, 354, 359, 362.
7. American Journal of Science, ser. 3, vol. 24 (124), 1882, p. 284-286.
8. American Journal of Science, ser. 3, vol. 26 (126), 1883, p. 271-294.
9. U. S. Geological Survey Bulletin 20, 1885, p. 40-68.
10. American Mineralogist, vol. 20, 1935, p. 322-326.
11. Colorado Geological Survey Bulletin 18, 1920, p. 55-57.
12. Rocks and Minerals, vol. 23, 1948, p. 206-207.

13. Mineralogist, vol. 9, 1941, p. 209-210.
14. Mineralogist, vol. 19, 1951, p. 283-286.
15. Mineral Resources, 1883-1884 (1885), p. 741.
16. U. S. Geological Survey Bulletin 207, 1902, p. 45-46.
17. Colorado Scientific Society Proceedings, vol. 14, 1945, p. 284-285.
18. Colorado Scientific Society Proceedings, vol. 15, 1949, p. 278-280.
19. American Philosophical Society Proceedings, vol. 16, 1877, p. 509-518.
20. American Journal of Science, ser. 3, vol. 44 (144), 1892, p. 385-386.
21. U. S. Geological Survey Folio 203, 1916.
22. Academie des sciences Comptes-rendus, vol. 109, 1889, p. 39-41.
23. American Mineralogist, vol. 42, 1957, p. 425-429.
24. American Journal of Science, ser. 3, vol. 19, 1880, p. 390-393.
25. Mineral Collector, vol. 1, 1895, p. 163-165.
26. U. S. Geological Survey Circular 220, 1952.
27. American Journal of Science, ser. 3, vol. 41, 1891, p. 439.
28. American Mineralogist, vol. 51, 1966, p. 299-323.
29. Proceedings of the Academy of Natural Sciences of Philadelphia, 1877, p. 9-11.
30. American Mineralogist, vol. 18, 1933, p. 115.
31. Geological Society of America Bulletin, vol. 62, 1951, p. 1517.
32. American Journal of Science, ser. 3, vol. 42 (142), 1891, p. 35-36.
33. American Mineralogist, vol. 33, 1948, p. 84-87.
34. American Journal of Science, ser. 4, vol. 7 (157), 1899, p. 51-57.
35. American Mineralogist, vol. 50, 1965, p. 1273.

MAPS

Topographic maps: Colorado Springs (two scales, 1948-61), Manitou Springs (1948-61), Mount Big Chief (two scales, 1948-61), Pikes Peak and Vicinity (1956, contour or shaded-relief editions, $1.00).
National forests: Pike.

COLORADO SPRINGS TO BURLINGTON

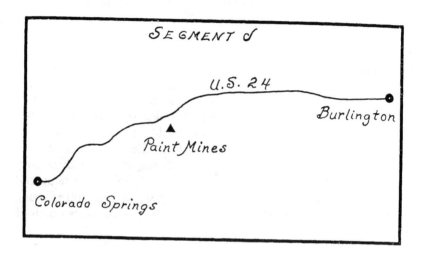

The highway between Colorado Springs and Burlington (near the Kansas border) crosses the Great Plains of central Colorado, through El Paso, Elbert, Lincoln, and Kit Carson Counties. The Paint Mines, near Calhan, 35 miles east of Colorado Springs, are described below. The characteristic minerals found in this region are petrified wood, agate, and jasper, all native to this section of the state, and minerals such as rock crystal, smoky quartz, and amazonstone, which have been carried by streams from the mountains on the west. These are widely scattered, and deposits of them are largely fortuitous. Local residents can often guide the collector to favorable collecting spots. Many new localities are awaiting discovery.

The Black Forest area, northeast of Colorado Springs, in El Paso County, yields abundant petrified wood. Peyton and Calhan, 25 and 35 miles, respectively, from Colorado Springs on U. S. 24, are centers for such specimens. The eastern part of El Paso County has always been a good source of petrified wood. In 1906, W. C. Hart collected and sold 5,000 pounds of it for gem and ornamental use. The Bijou Basin, a conspicuously eroded topographic depression in Elbert County – reached from Peyton – is noted for its brown jasper and petrified wood. A short distance southeast of Elbert, in Elbert County, logs up to 2 feet in diameter and 25 feet across have recently been found, some of them being well opalized and suitable for cutting. Eugene M. Beason (Ref. 1) described and mapped a few of the many specific localities in the Peyton-Elbert area, suggesting that the washes and ravines are the best places to look. Lucile and Elmer Brown (Ref. 2) offer permits to collect on a ranch near Simla, in Elbert County, to persons inquiring at the Lucy Brown Rock and Gift Shop. The village of Agate, also in Elbert County, is well named. Petrified, agatized, jasperized, and opalized wood occur in eastern Colorado in a zone about 150 miles wide, as outlined by Theodore H. Kleeman (Ref. 2). An opalized log more than 45 feet long is displayed in the University of Colorado Museum, in Boulder. The 1965 floods uncovered much new material. Considerable moss opal has been found along the South Fork of the Republican River, about 20 miles north of Burlington.

REFERENCES
1. Gems and Minerals, no. 359, 1967, p. 26-27.
2. Gems and Minerals, no. 385, 1969, p. 48.
3. Mineralogist, vol. 9, 1941, p. 253.

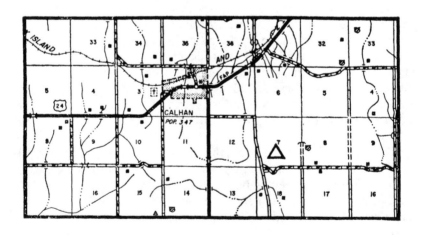

The Index to Topographic Maps of Colorado, obtainable free from the U. S. Geological Survey, Federal Center, Denver, Colorado 80225, shows which parts of east-central Colorado are covered by topographic maps.

Paint Mines

Attractive specimens of selenite (the clear, crystalline variety of gypsum) and some colorful varieties of jasper are found in the sedimentary rocks at the locality known as the Paint Mines, near Calhan, in northeastern El Paso County. This is one of the most attractive small scenic spots in eastern Colorado – a miniature combination of the Bad Lands and Bryce Canyon, which would be worth a visit even if no minerals were present to be collected. The various gullies, however, are somewhat less than 100 feet deep. The grotesque erosional features, called hoodoos, consist of white, cross-bedded sandstone, which gleams in the sunlight, alternating with richly colored shale and clay of many hues, especially bright purple, yellow, and green. Caps of sandstone are curiously supported by pointed pillars of clay, which are thereby temporarily protected from destruction. This locality was first made known to the author by Mrs. Robert Hymen, of Colorado Springs, and Clarence G. Coil, of Colorado Springs, first referred him to the selenite to be found here. The location is on private ranch land owned by Harry Freeman and others, who have not hitherto restricted visitors, as the fence nearest the main gullies is not closed.

The log from Calhan is as follows:

0.0 Intersection of U. S. 24 and main street corner in Calhan. Go east on U. S. 24.
0.4 Turn right past filling station onto asphalt road.
1.0 Turn left at junction onto gravel Paint Mine Road.
3.0 Fork left.
3.2 Turn left through fence and descend into main gullies.
3.4 Park near edges of gullies.

The rock consists of the Dawson Formation, of early Tertiary age. The selenite is enclosed in the black-shale ledges at the far (north) end of the canyon. These may perhaps be better reached earlier from the gravel road, 0.2 miles before the road fork mentioned above. The jasper has been picked up on the surface of the ground above the gullies.

According to Harry W. Oborne, of Colorado Springs, there are a number of other groups of gullies in the vicinity. A particularly scenic area, with prominent white monoliths, is situated south of the dump piles near the same gravel road. The Plains Indians of the Pikes Peak region and the Van Briggle Art Pottery, of Colorado Springs, are reported to have taken colored clay from this area in years past. Occasional Indian artifacts are picked up here.

REFERENCES

1. Mineral Notes and News, no. 163, April 1951, p. 11.

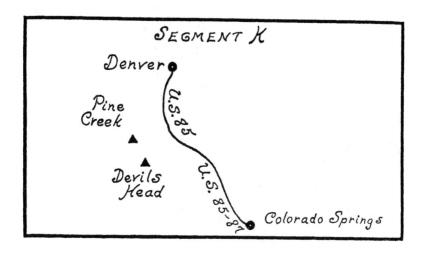

The highway from Colorado Springs to Denver follows the foothills of the Front Range, presenting occasional, spectacular views of the main mountains. The localities of Devils Head and Pine Creek, both described below and within the area covered by the Pikes Peak granite, are reached by going west from this road, which passes through El Paso, Douglas, Arapahoe, and Denver Counties.

Devils Head

Cavities in pegmatite in the Pikes Peak granite at Devils Head in western Douglas County have yielded the finest gem topaz ever found in Colorado. The largest crystal – perhaps the largest well-formed and well-preserved topaz crystal ever found in North America (although larger cavernous crystals have come from Maine) – weighed 1,160 grams (37.1 ounces) and was found in 1934; it was a rich, reddish brown. A symmetric crystal weighing 587 grams (18.8 ounces) was found here in 1885 and was illustrated (through wrongly labeled) in George F. Kunz's noted book

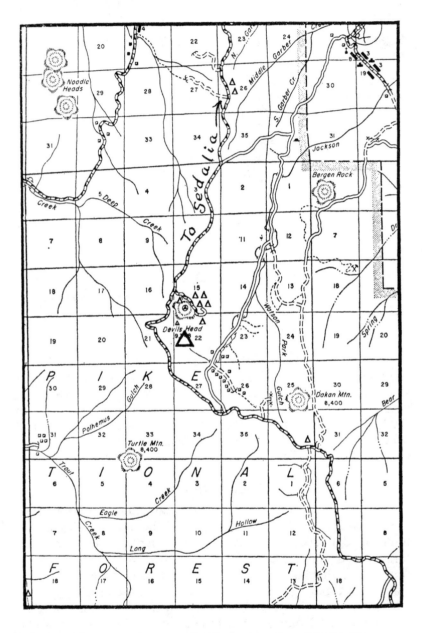

Gems and Precious Stones of North America (1890). The complete transparency and often the delicate color of many of the crystals, although most of the surfaces are rough and stained, have made possible the cutting of beautiful gems. The topaz is colorless, light yellow, cinnamon brown, reddish brown, and bluish. A sherry stone, cut to 235 carats, in the U. S. National Museum, faded completely after exposure to daylight.

Besides topaz, some of the largest crystals of smoky quartz in Colorado and abundant specimens of rock crystal have come from Devils Head. Several kinds of feldspar are common. In order to maintain noncommercial collecting rights for all individuals, the Colorado Mineral Society in 1957 filed a mining claim upon the most noted of the Devils Head workings. Lawrence Oliver and C. W. Hayward, both of Denver, were especially instrumental in staking this claim.

Reaching an altitude of 9,348 feet, Devils Head is a large, rather isolated peak. In the classic atlas of the Hayden Survey, it was called Platte Mountain. It is drained by Jacksons Creek and by intermittent tributaries of the northward-flowing South Platte River, about 6 miles to the west.

The Devils Head area can be reached on the Rampart Range Road, which follows the crest of the range northward from the Garden of the Gods, at Colorado Springs, or from Woodland Park. This road should not be attempted except in summer, owing to heavy show that presents considerable hazard. From a branch of the main Colorado Springs-Denver highway (U. S. 85), the approach is from Sedalia, according to the following log:

0.0 Johnson's Corner, at Sedalia. Go west on Jarre Canyon Road, Colo. 67.

10.2 Turn left on Rampart Range Road, gravel. (Road ahead goes to Pine Creek locality, described below).

16.7 Keep right at junction.

19.3 Keep right at fork. (Left road goes to Devils Head Campground).

20.4 Turn left at C.M.S. Topaz Claim sign on tree across road. Do not try to park more than one car at the bottom. Park part way down or go to the Virgin's Bath parking (see below) and walk back to the deposits.

20.8 End of road. Claim is here; other prospects are evident in this area.

The original discovery, first mentioned in 1883 by R. T. Cross (Ref. 1), was made by Walter B. Smith, who described the locality soon afterward (Ref. 2). Large-scale work was done in 1934 by Edwin W. Over, Jr., and Arthur Montgomery; some of the results of their finds were described by M. A. Peacock (Ref. 3). James B. Craig, of Denver, has taken huge crystals of smoky quartz from ground adjacent to the Colorado Mineral Society claims in recent years.

The topaz crystals at Devils Head have been found lying loose in cavities in the Pikes Peak granite; also in clay and sand, stained brown by iron; and in gravel eroded downslope. The solid parts of the pegmatite consist of the usual quartz and microcline feldspar, some very coarse, together with masses of worthless topaz as large as one's fist. Shoveling the surface rock is easy, because it is so decomposed and disintegrated. Exposing the pockets themselves, however, has entailed extensive blasting.

The rock was obviously broken by movement after the formation of the minerals in the cavities. Alteration and etching of the topaz is common. Iron-bearing solutions have penetrated the many fissures in the rock, and iron oxides have been deposited along the basal cleavage cracks in the topaz crystals. The most productive pockets seem to be those with the least iron.

Other minerals have been found with the topaz, and some may be found independently. Microcline feldspar is the most common, both in masses up to about 8 inches across and in crystals a couple of inches long. The green (amazonstone) variety, however, is greatly inferior to that from some other places in the Pikes Peak region. Albite feldspar occurs in rounded groups. The various feldspars have partly altered to kaolinite and also make up most of the sand in the pockets.

Smoky quartz is especially common, mostly in narrow crystals, some bearing upon them small, flaky crystals of goethite and hematite. There is much rock crystal, most of it stained by iron, although many fragments are brilliantly clear, especially when found near the best topaz. Amethyst has been seen by Glenn R. Scott. Muscovite mica occurs in discolored lumps and flakes of sericite. Cassiterite has been found very rarely with the topaz in albite or quartz. Fluorite crystals occur in poor combinations of octahedron and cubes. Glossy, pitch-black crystals of allanite associated with dark-green and black gadolinite; the description in 1886 by L. G. Eakins (Ref. 4) was the first report of gadolinite in the United States. Associated with these last two minerals are cyrtolite (a radioactive variety of zircon) and samarskite, as described by W. F. Hillebrand (Ref. 5-6).

Continuing log:
20.8 Road to Colorado Mineral Society claim mentioned above. Continue.
20.9 Virgin's Bath Scenic Point parking place. The shallow ravine across the road on the left has often led to good collecting places.
21.4 White Quartz Mountain area.

On the south slopes, for a distance extending ¼ to ½ mile west of the highway, can be found quartz crystals of the clear and smoky varieties. Both are also picked up in the stream beds and gullies.

Continuing log:
20.9 Virgin's Bath Scenic Point parking place mentioned above. Continue.
26.2 Turn right into Long Hollow on Bergen Road.

A 62-pound crystal of smoky quartz and several weighing about 45 pounds were collected in Long Hollow about 1968 by Frank Kegler, of El Paso County. A 44-pound crystal was uncovered here in 1944 by Louis and Grace Binderup, of Denver. At the same locality in 1953, Algird Stephan, of Denver, discovered a crystal of topaz which weighed 17½ pounds and measured nearly 12 inches in length and 4 to 5 inches in width. James R. Hurlbut, of Denver, reports the presence of similarly large topaz crystals still embedded in the rocks at Long Hollow. Glenn R. Scott has reported crystals of bluish-green, sugary-textured topaz up to 2 feet long and 8 inches across. It is the author's opinion that this general area of Devils Head offers more long-range possibilities for mineral collecting than any other place in Colorado.

REFERENCES

1. American Journal of Science, ser. 3, vol. 26 (126), 1883, p. 484-485.
2. U. S. Geological Survey Bulletin 20, 1885, p. 72-74.
3. American Mineralogist, vol. 20, 1935, p. 354-363.
4. Colorado Scientific Society Proceedings, vol. 2, 1886, p. 32-35.

5. Colorado Scientific Society Proceedings, vol. 3, 1888 (1889), p. 38-45.
6. U. S. Geological Survey Bulletin 55, 1889, p. 48-52.
7. Mineralogist, vol. 9, 1941, p. 416, 418-419.
8. U. S. Geological Survey Professional Paper 227, 1950, p. 31.
9. U. S. Geological Survey Bulletin 1114, 1961, p. 18, 37, 91, 122, 140, 142, 154, 276, 290, 333-334.
10. Gems and Minerals, no. 309, 1963, p. 16-19.

MAPS

Topographic maps: Dakan Mountain (1956), Devils Head (1954). National forests: Pike.

Pine Creek

The Pine Creek locality is a rather indefinite one in Douglas County, extending for some little distance. Many old mine dumps, some covered by vegetation, are to be seen in this area, and the steep slopes are heavily forested. Good specimens of rock crystal, smoky quartz, and amazonstone are still found on the dumps, and unopened pegmatites are numerous. Profitable collecting may be done by digging in loose dirt, especially around the roots of trees.

The area is reached from Sedalia according to the following log:

0.0 Johnson's Corner at Sedalia on U. S. 85. Go west on Jarre Canyon Road, Colo. 67.

10.2 Pass Rampart Range Road to Devils Head, described above. Gravel road ahead.

13.7 Keep right at Sprucewood Inn, formerly Pine Creek Store. (Left-hand road, Colo. 67, goes to Deckers.) After about 2 miles on this steep road, dumps, a tunnel, and several pegmatite dikes can be seen across the gulch on the left, above the road.

Charles W. Reitsch (Ref. 1) has described crystals of smoky quartz up to 6 by 12 inches. Considerable gem material is present, especially in the smaller crystals; even the larger ones may be as much as half gemmy. The smoky color is usually rather pale. Most of the quartz is malformed and iron stained.

Some gem amazonstone is found here. It grades into common

microcline feldspar, mostly cream to flesh colored. Albite feldspar occurs in radiating groups of crystals. The cleavage cracks in the feldspar are stained brown, and the specimens crumble easily. Numerous pieces of feldspar have short crystals on top of them, and there are various intergrowths of the two minerals. Embedded in some of the feldspar are cubes and irregular masses of amber to yellow fluorite, some crystals being as much as 3 inches across, clear inside but rough and stained outside. Cassiterite has been reported by Glenn R. Scott to Edwin B. Eckel (Ref. 2).

REFERENCES

1. Rocks and Minerals, vol. 14, 1939, p. 270-271.
2. U. S. Geological Survey Bulletin 1114, 1961, p. 18, 91, 140, 142, 150-151, 276.
3. Lapidary Journal, vol. 23, 1969, p. 1016-1019.

MAPS

Topographic maps: Deckers (1954).
National forests: Arapaho, Pike.

DENVER MOUNTAIN PARKS

The Denver Mountain Parks System provides a mineral-collecting area that is unique in the country. Dozens of small mines and quarries in the Foothills and the Front Range are accessible by highway, within a relatively short distance of the capital city. This municipally owned area contains about 25,000 acres in 51 separate tracts of land. Of these, 28 are named, and they range in size up to Genesee Park's 2,341 acres. The highest spot, Summit Lake, is at an altitude of 12,740 feet. The best known place is the Park of the Red Rocks (with its Red Rocks Amphitheater), near where occur some of the most prolific dinosaur beds in the world, as described in *Exploring Rocks, Minerals, Fossils in Colorado*, by Richard M. Pearl (Sage Books, revised edition, 1969). The connecting roads include the highest automobile highway in America, reaching nearly to the top of 14,260-foot Mount Evans.

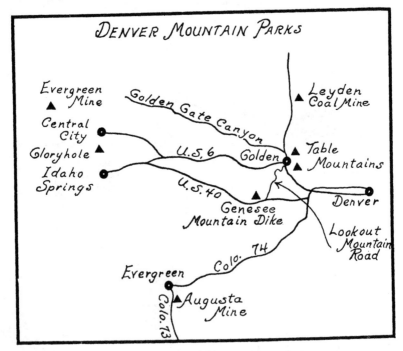

The logs given below are designed to reach the best recommended mineral deposits in the most convenient order, rather than to take in the major tourist attractions. Each log begins at a nearby, easily identified place, to avoid running up a large total mileage that might introduce speedometer errors.

Although the Denver Mountain Parks quadrangle of the U. S. Geological Survey encloses all the parks (as well as the many mines of Idaho Springs), this section of the book has been enlarged a bit to take in the convenient localities in and near Golden Gate Canyon, and also the Leyden coal mine, as well as including the Table Mountains (adjacent to Golden) and the localities near Central City.

Owing partly to indecision and partly to unfamiliarity with some of the area, this section of the state was rather neglected in previous editions of this book and its predecessor. In the summer of 1964, dozens of localities in this area were examined — including the many enumerated by Margaret Fuller Boos (Ref. 1) — and it is apparent that little of major value has been missed by this oversight. Nevertheless, the nearness to Denver seems to justify giving the area more attention. The Table Mountains and Gloryhole localities, however, have always been of outstanding interest.

The various mining towns situated directly west of Denver are quaint, historic, and richly mineralized. The first important discovery of placer gold in Colorado was made on January 7, 1859, on the outskirts of Idaho Springs. Gold was discovered in veins on May 6 of the same year, at a site marked by a monument at the boundary line between Blackhawk and Central City. Two other important mining towns that are adjacent to each other are Georgetown and Silver Plume, only 2 miles apart. Empire is 5 miles north of Georgetown. These localities were famous chiefly for gold — silver, lead, zinc, and copper being of lesser importance. The pitchblende deposits at Central City came back briefly into prominence. Ore and gangue specimens are abundant on the many mine dumps but are mostly of little value, being pyrite, galena, sphalerite, chalcopyrite, and quartz. The geology and mineralogy of this region have been summarized by John W. Vanderwilt (Ref. 2). The geology and ore deposits of the Front Range were covered in detail by T. S. Lovering and E. N. Goddard (Ref. 3).

Mineral collectors visiting Denver are invited to attend the meetings of the Colorado Mineral Society, which are held on the first Friday of every month from October through May, at the Jefferson County Fairgrounds, 15,200 West 6th Avenue.

The Denver Gem and Mineral Guild meets on the second Friday of the month at the First Federal Savings and Loan Building, 3400 West 38th Avenue.

The Gates Rock and Mineral Club meets on the second Tuesday of each month from September to May at the Gates Rubber Company, 999 South Broadway, Denver.

The Littleton Gem and Mineral Club meets on the last Friday of the month at the Colorado Public Service Building, Sheilane.

In Arvada, the North Jeffco Gem and Mineral Club meets on the second Friday of the month at the North Jeffco Park and Recreation District Administration Building, 9101 Ralston Road.

In Westminister, the Mile Hi Rock and Mineral Society meets on the fourth Friday of the month from September to May at the Advent Lutheran Church, West 80th Avenue and Meade Street.

The Denver Museum of Natural History is a leader in the world of museums. Besides its magnificent mounted skeletons of dinosaurs and other prehistoric beasts, its wonderfully realistic animal-habitat groups, and its other exhibits, there are on display rocks, minerals of exceptional beauty, and fine gems. The hours are 10:00 to 4:30 on Monday to Saturday, and 12:00 to 5:00 on Sunday and holidays; summer hours are 9:00 to 5:00 every day.

The Geologic Museum of the Colorado School of Mines, in Berthoud Hall, Illinois and 16th Streets, on the campus in Golden, is one of the most modern and handsomely equipped in the country. Exceptionally fine collections of about 3,500 minerals and rocks and 1,000 fossils are on display from 8:00 to 5:00 on Monday to Friday and 1:30 to 4:00 on Saturday and Sunday. The Frank C. Allison safe of gold specimens from Farncomb Hill in the Arthur Lakes Library is also well worth seeing; the building is open Monday to Friday from 8:00 to 5:00.

The Colorado School of Mines acquired in 1966 the superb mineral collection formerly maintained by the State Historical Society, and before that by the State Bureau of Mines, in the State Museum Building across from the State Capitol, in Denver. It will someday be placed on exhibit again. A chief feature of the collection, which numbered 11,177 items, was its arrangement by counties, so that miners, prospectors, and collectors could learn which minerals have been found in each county. The State Historical Society retains a representative collection of Colorado minerals among its historical displays.

REFERENCES

1. Geological Society of America Bulletin, vol. 65, 1954, p. 115-142.
2. *Mineral Resources of Colorado*, State Mineral Resources Board, Denver, 1947.
3. U. S. Geological Survey Professional Paper 223, 1950.

MAPS

Topographic maps: Central City (two scales; 1903-10, 1942), Conifer (1957-65), Denver Mountain Area (1938-48; contour or shaded-relief edition, $1.00), Denver Mountain Parks (1903-23; $1.00), Evergreen (1957-65), Georgetown (two scales, 1957), Golden (1957-65), Harris Park (1957), Idaho Springs (1957), Indian Hills (1957-66), Meridan Hill (1957), Morrison (1957-65), Mt. Evans (1957), Ralston Buttes (1965), Squaw Pass (1957). National forests: Arapaho, Pike.

Table Mountains

Two basalt-lava mesas just outside the college city of Golden, in Jefferson County, constitute one of the most important mineral localities in Colorado and a source of zeolites that is world famous. Standing guard about 700 feet above a narrow valley, North Table Mountain and South Table Mountain, one on each side of Clear Creek, tempt the mineral collector with superb specimens of a dozen different zeolites — many of them outstanding for their exquisite beauty and richness of form — in addition to apophyllite, aragonite, halloysite, and luminescent calcite. Every new exploration suggests that neither the supply nor the quality has diminished during the 90 years since this locality was first mentioned in print.

North and South Table Mountains, having maximum altitudes of 6,570 feet and 6,319 feet, respectively, were once part of a single body of volcanic rock until the stream cut a sharp gorge between them. They were formed by three successive flows of lava, which spread from a vent believed to have been situated

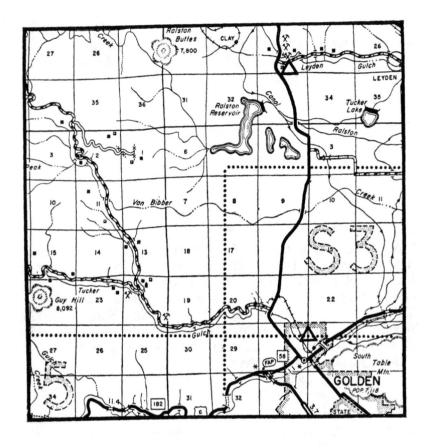

under the west-central part of North Table Mountain or else at Ralston dike, somewhat to the north. This lava cooled and solidified to become basalt and basalt porphyry; in the fissures and gas cavities were deposited the zeolites and other minerals of late origin.

The pioneer basic geology of the area was reported by Samuel M. Emmons, Whitman Cross, and George H. Eldridge (Ref. 1), and the first detailed description of the mineralogy was given by Whitman Cross and William F. Hillebrand (Ref. 2-3). The most essential additions, corrections, and refinements have been provided by the studies of Horace B. Patton (Ref. 4), J. Harlan Johnson (Ref. 5-6), William A. Waldschmidt (Ref. 7-8), Leslie W. Leroy (Ref. 9), and Stanley O. Reichert (Ref. 10). An excellent summary appears in the text matter of Geologic Quandrangle Map

CQ 103, by Richard Van Horn, issued in 1957 by the U. S. Geological Survey (Ref. 11).

Three lava flows are present on North Table Mountain. On South Table Mountain, only the two capping layers exist; they are directly on top of each other and are separated from the lowest layer by 150 feet of shale and sandstone. The sides of the three flows are distinguishable as cliffs, while the tops are marked by benches in the profiles of the mesas, as seen best from the valley approaching Golden.

This is an important feature to the collector, inasmuch as the zeolites occur predominantly in the less resistant upper part of the middle flow, a zone up to 25 feet thick, distinguished by a porous (vesicular or scoriaceous) texture and a reddish or brown color at its contact with the uppermost layer. Numerous zeolites also occur immediately beneath this (in the same flow) in a zone recognized by vertical columnar jointing. Some zeolites have been observed in the topmost flow, but none whatsoever in the bottom flow.

The superior hardness of the jointed basalt cliffs has protected the softer underlying sedimentary rock layers from erosion. The steep, lower slopes of the mesas are covered with fragments of broken basalt, some of them peculiarly rounded.

The lava flows belong to the Denver Formation, deposited shortly after the creation of the present Rocky Mountains, close to 60 million years ago. Associated with this formation are found considerable petrified palm wood, fossil leaves, and bones of dinosaurs, crocodiles, and primitive mammals.

Many quarries have been operated on all sides of both mesas since the earliest settlement in the region. Paving blocks and surfacing for Denver streets, flood-control walls for the South Platte River and Cherry Creek, railroad ballast, concrete aggregate, monument stones (because of its ability to take a high polish), and water and sewage treatment are among the uses to which the basalt has been put. Enormous quantities have been extracted since the Second World War, especially from 1949 to 1951, for the Cherry Creek and Harlan County (Nebraska) Dams, the latter using over half a million tons of the rock in making concrete (Ref. 12).

Such active mining continually exposes new sources of zeolites and destroys old ones. The chief problem for the mineral collector, superseding the former one of breaking into new pockets, has therefore become that of securing access to the workings before it is too late. The top of North Table Mountain is the property of F. A. Foss, who owns and operates the Walgreen drugstore in downtown Golden; he lives beneath the mesa and has readily granted permission to enter the fenced enclosure when requested. South Table Mountain is used by several ranchers for grazing and pasturing. The lower slopes of the mesas are being developed for residential use. A warning against rattlesnakes is necessary.

The gas cavities, which are enlongated in a horizontal direction, are frequently as much as 1 foot in diameter — rarely up to 8 feet — and the smaller ones, down to a fraction of an inch, are apt to be completely filled with minerals. Such an open space, called a vesicle, is termed an amygdule when it is lined or filled by later mineralization; the rock here is thus an amygdaloidal basalt. In some of the larger cavities, the zeolites have been deposited on the floor in distinct layers, indicating intervals between deposition. The minerals are usually quite fresh and intact when newly exposed to a depth of a couple of feet.

Guy B. Ellermeier, of Denver, has given useful suggestions (Ref. 13) for collecting at this locality. He believes the use of a large cold-chisel and a heavy hammer to be sufficient, in addition to a diamond-pointed tool and a square-pointed one available for extracting specimens from the rock. He also finds that the largest specimens are obtained by blasting a big boulder, and that cutting is facilitated by keeping the basalt wet ahead of the diamond point.

Zeolites are a related group of minerals characterized by their easy fusibility and their property of yielding water continuously when heated. At the Table Mountains locality, the most common of the zeolites are thomsonite (first), analcime (second), natrolite, mesolite, and chabazite. Also found are heulandite, laumontite, levynite, nordenite (earlier called ptilolite), scolecite, and stilbite.

The following descriptions apply to the more abundant specimens; some of the zeolites here (especially thomsonite) are remarkable for their great variety of types and may appear in surprising habits, each representing a different generation, totaling altogether at least 15 periods of deposition. A visit to the nearby Geologic Museum in Berthoud Hall of the Colorado School of Mines is advisable before attempting to collect on the mesas, in

198

order to become acquainted with the appearance and the names of the minerals. For more complete descriptions, the important original publications (Ref. 2-7, 14), the article by Mr. Ellermeier (Ref. 13), and the summaries by Edwin B. Eckel (Ref. 15) are recommended.

Thomsonite is most typically found in bunches of radiating, fibrous, colorless crystals, which often assume a globular aspect. Analcime occurs as trapezohedrons, the smaller ones being transparent and colorless, the larger ones (ranging up to 2½ inches in size) being white and opaque. The crystals of natrolite are needle-like, being long, slender, transparent, and colorless; some natrolite occurs as mealy crusts. Mesolite usually resembles gauze, felt, or tufts of delicately spun cotton. Chabazite grows in single or twinned rhombohedrons and other forms and has a white or pinkish color.

Associated with the zeolites is apophyllite, a closely related mineral, present in small, rectangular, white crystals with terminal faces unless broken across; a greenish tinge is fairly noticeable. Also present are snow-white aragonite, as well as yellow and colorless calcite of the dog-tooth variety. Most of the calcite is fluorescent, some is phosphorescent, and some is thermoluminescent, meaning that it glows (orange) in the dark when warmed. Halloysite, a clay mineral, is also present. Wavellite was reported in 1878 by F. M. Endlich (Ref. 16) but not since, and so it was probably in error.

Arthur Lakes (Ref. 17) once stated that analcime, thomsonite, and natrolite from this locality were sometimes cut for small gems.

REFERENCES

1. U. S. Geological Survey Monograph 27, 1896, p. 292-296.
2. American Journal of Science, ser. 3, vol. 23 (123), 1882, p. 452-458.
3. American Journal of Science, ser. 3, vol. 24 (124), 1882, p. 129-138.
4. Geological Society of America Bulletin 11, 1900, p. 461-474.
5. American Mineralogist, vol. 10, 1925, p. 118-120.
6. Colorado School of Mines Quarterly, vol. 25, no. 3, 1930.
7. Colorado School of Mines Quarterly, vol. 34, no. 3, 1939.
8. University of Colorado Studies, vol. 26, 1938, p. 143-146.

9. Colorado School of Mines Quarterly, vol. 41, no. 2, 1946.
10. Colorado School of Mines Quarterly, vol. 49, no. 1, 1954.
11. U. S. Geological Survey Geologic Quandrangle Map GQ 103, 1957.
12. Colorado School of Mines Quarterly, vol. 44, no. 2, 1949, p. 116-127.
13. Rocks and Minerals, vol. 22, 1947, p. 618-623.
14. Colorado School of Mines Quarterly, vol. 33, no. 3, 1938.
15. U. S. Geological Survey Bulletin 1114, 1961, p. 24, 50, 56, 86, 91, 178, 185, 202, 205, 221, 234-235, 240, 295, 312, 330, 354.
16. Tenth Annual Report of the United States Geological and Geographical Survey of the Territories . . . 1876 (1878), p. 133-159.
17. Mining World, vol. 30, 1909, p. 831-832.
18. U. S. Geological Survey Bulletin 207, 1902, p. 34-35, 45-46.
19. American Mineralogist, vol. 18, 1933, p. 402-406.
20. Mineralogical Magazine, vol. 33, 1932, p. 54.
21. Rocks and Minerals, vol. 11, 1936, p. 52.
22. Annual Report of the United States Geological and Geographical Survey of the Territories . . . 1874 (1876), p. 129-130, 136.
23. American Journal of Science, ser. 4, vol. 9 (159), 1900, p. 117-124.
24. Neues Jahrbuch, vol. 1, 1884, p. 250-256.
25. U. S. Geological Survey Bulletin 20, 1885, p. 13-39.

MAPS

Topographic maps: Denver Mountain Area (1938-48; contour or shaded-relief edition, $1.00), Golden (1957-65), Morrison (1957-65).
National forests: Arapaho, Pike, Roosevelt.

Leyden Coal Mine

Mineral collectors who must have pretty specimens might not be interested in this locality. It is nevertheless, one of the easiest places in the country to obtain a few uranium minerals, and it is a historic spot. From here, the important uranium mineral carnotite was first described as early as 1875 (by E. L. Berthoud, Ref. 1), although it was not until 1889 that it was identified and named (from material from Montrose County). In 1904, the Leyden mineral was proved to have been the same, according to Herman Fleck (Ref. 2). Specimens from here were exhibited at the Centennial Exposition in Philadelphia in 1876.

The carnotite is present as a yellow coating in coal of the Laramie Formation, of Cretaceous age. The newer uranium mineral coffinite occurs here as a black coating, which may be mistaken for the coal itself. A Geiger counter will be helpful in selecting specimens. The coal is partly silicified and contains vugs of quartz, as described by John H. Wilson (Ref. 3). The uranium minerals were deposited in a bed of permeable sandstone, separated from the actual coal by a layer of claystone, according to Eugene L. Grossman.

The Leyden mine is a flat opencut along the edge of a conspicuous hogback cut by Leyden Creek. The land was originally homesteaded in 1869 by Patrick Stanton. A former shaft has caved in. The Public Service Company of Colorado is now using the mine for the underground storage of natural gas.

This locality can be reached from Golden by way of the road to Boulder, Colo. 93. From the welcome arch over Washington Street in downtown Golden, the log is as follows:

0.0 Go north on Washington Street, Colo. 93.
7.3 Turn right on gravel road toward Leyden, passing through a break in the hogback.
7.5 Park at mine entrance.

REFERENCES

1. Proceedings of the Academy of Sciences of Philadelphia, vol. 27, 1875, p. 363-365.
2. Colorado Scientific Society Proceedings, vol. 11, 1916, p. 155-156.
3. Engineering and Mining Journal-Press, vol. 116, 1923, p. 239-240.
4. U. S. Geological Survey Bulletin 1114, 1961, p. 23, 89-90, 110.

MAPS

Topographic maps: Golden (1957-65).
National forests: Arapaho, Roosevelt.

Golden Gate Canyon

Two minor but accessible localities are situated along the Golden Gate Canyon road, Colo. 58, and might prove interesting to mineral collectors who are in the vicinity of Golden or who plan to travel on this road between Golden and Central City or to the newly established Golden Gate Canyon State Park. Several other, more important localities are also situated along this road but are not accessible without special permission from the owners. These include the well-known Centennial Cone pegmatite, Drew Hill pegmatite, and Guy Hill (Ramstetter Ranch) pegmatite; the first two were described in *Colorado Gem Trails*, but had to be omitted later because they were closed to the general collector. Many pegmatites appear along the road, mostly in metamorphic rock.

No Name Pegmatite

Boulders of pegmatite containing black tourmaline, large pieces of muscovite mica, pink feldspar, and quartz occur conveniently along both sides of the road at this locality in Golden Gate Canyon. Tourmaline has been found widely in this general area since the earliest days; some crystals have been as large as 1 foot in length (see Edwin B. Eckel, Ref. 1). Except on fenced property, further search should be encouraged.

The locality can be reached from Golden by the following log:

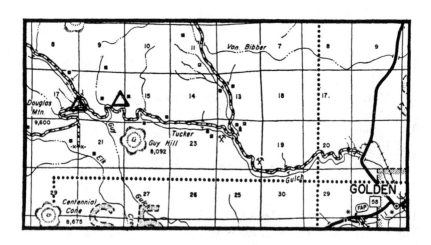

0.0 Welcome arch across Washington Street in downtown Golden. Go north on Washington street.

1.5 Turn left onto Golden Gate Canyon Road, Colo. 58 (gravel), avoiding side road straight ahead.

1.6 Pass Pine Ridge Road on right. Pass side roads, bronze monument, grange, old school, and various houses.

5.6 Pass Crawford Gulch Road on right.

8.5 Turn right onto locality opposite road and mailboxes on left.

REFERENCES

1. U. S. Geological Survey Bulletin 1114, 1961, p. 337-338.

MAPS

Topographic maps: Ralson Buttes (1965).
National forests: Arapaho, Pike, Roosevelt.

Robinson Gulch Prospect

The Robinson Gulch locality yields chrysoberyl and black tourmaline, the former of good quality, but the supply is limited. A pegmatite dike consisting mostly of white feldspar stands by the side of the road.

Continuing log:

8.5 Turnoff mentioned above. Continue on Colo. 58.

9.9 Park at junction of Robinson Hill Road and walk 20 feet up road. Pegmatite on right. A good picnic spot is across Guy Gulch before the junction.

REFERENCES

1. U. S. Geological Survey Bulletin 1114, 1961, p. 23-24, 106.

MAPS

Topographic maps: Ralston Buttes (1965).
National forests: Arapaho, Pike.

Genessee Mountain Dike

Small amounts of several interesting minerals can be obtained from this pegmatite body. This is not a spectacular locality, but it is situated in an accessible place, and there is always the chance of finding more specimens. The garnet has been of remarkable size. The log to the Genesse dike is as follows:

0.0 Junction of Int. 70-U. S. 40 and Lookout Mountain Road. Go west on Int. 70.

0.6 Stop off road near Conoco station and store. Walk back as far as curve in road and private road opposite. Take care along this heavily traveled highway.

The minerals are in a large pink-granite pegmatite, a quartz-garnet dike, and the metamorphic rock of the Idaho Springs Formation, which they intersect. It has been suggested that the quartz-garnet dike is an offshoot of the pegmatite from below the road surface. This deposit was described in a Colorado School of Mines thesis by Claude Utterback and Kenneth Hueckel (1941). Additional minerals have been reported to Edwin B. Eckel (Ref. 1) by Glenn R. Scott and John W. Adams.

The outstanding mineral is grossularite garnet, in fine crystals up to nearly 12 inches in diameter. Much broken, bright-brown garnet is in evidence. Sphene occurs in sharp, easily recognized, dark-brown crystals as much as 1 inch across. Scheelite is in rough crystals, sometimes embedded in the garnet. Brown allanite, brown astrophyllite, yellowish-green epidote, black hornblende, and hyalite opal are also found in and around the dike; wernerite (in cavities in hornblende) and greenish-blue idocrase have also been reported. The pegmatite contains some magnetite and vermiculite.

REFERENCES

1. U. S. Geological Survey Bulletin 1114, 1961, p. 23, 37, 162, 292, 293, 308, 350.

MAPS

Topographic maps: Evergreen (1957-65).
National forests: Arapaho, Pike.

Augusta Mine

Soft masses of deep-purple fluorite, as well as cleavage fragments that are green and colorless, occur at the Augusta mine. Small crystals of willemite have been reported in one of the few occurrences of this mineral in North America. Small amounts of yellow sphalerite, galena, and oxidized copper minerals (chalcocite, malachite, azurite) may still be found here, along with cerussite, barite, and limonite.

The log of this mine from Evergreen is as follows:

0.0 Bridge at intersection of Colo. 74 and Colo. 73. Go on Colo. 73 toward Conifer.

1.5 Pass side road and immediately turn left onto mine property, now used as a dump.

The former underground workings, which included a shaft 130 feet deep, are entirely inaccessible. The best material is found in the main vein in back of the isolated block of bedrock above the old tunnel. The crustified, quartz-fluorite vein ranges from 1 to 5 feet in width and averages 2 feet. The fluorite breaks into numerous stringers within the vein. Small crystals of black willemite were found in pockets of green fluorite by Jack W. Adams and Richard V. Gaines, as described by Frederick H. Pough (Ref. 1). Masses of copper, zinc, and lead ore (containing silver), for which the mine was originally worked, weighed up to 50 pounds. A

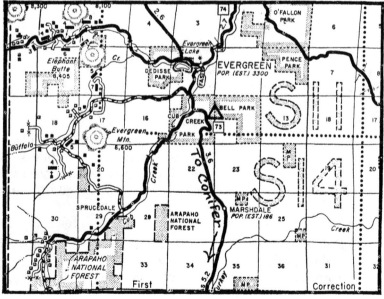

narrower vein parallels the main one about 100 feet to the east (left). The surrounding rock is hornblende-biotite gneiss and red granite.

Fluorite was mined here in the 1870's for use as a flux in the smelting of gold and silver at Central City. Later, over 1,000 tons of fluorite was marketed in Pueblo. The property has been enlarged by the recent removal of gravel for road construction.

Other exposures, probably of the same Augusta vein, are known on the northwest side of Cub Creek along Colo. 334, about ¾ mile southwest of Evergreen, and along Colo. 98, ¼ mile west of Colo. 74. Through openings along its course, the vein can be traced several miles to the south, as described by Harry A. Aurand (Ref. 2).

REFERENCES

1. American Mineralogist, vol. 26, 1941, p. 92-102.
2. Colorado Geological Survey Bulletin 18, 1920, p. 58-60.
3. U. S. Geological Survey Bulletin 340, 1908, p. 170.
4. U. S. Geological Survey Professional Paper 223, 1950, p. 279.
5. Geological Society of America Bulletin, vol. 26, 1915, p. 84.
6. Mining and Scientific Press, vol. 99, 1909, p. 258-261.
7. U. S. Geological Survey Bulletin 1114, 1961, p. 22, 100, 151, 355.

MAPS

Topographic maps: Conifer (1957-65).
National forests: Arapaho, Pike.

The Gloryhole

Surrounding the enormous openpit known as the Gloryhole, or the Patch, situated on the crest of Quartz Hill, southwest of Central City, are found small but doubly terminated crystals of quartz and attractive, small crystals of pyrite. Pyrite and magnetite also occur in several nearby places, as described here.

The log to these localities from Central City is as follows:

0.0 Main junction near Teller House. Go on Colo. 279 toward Idaho Springs.
1.5 Turn left downward before crest of road and descend past

stone foundations of old mills toward ghost mining town of Russell Gulch. Keep on right road.

2.1 Park off road. Pyrite on large dump on right slope, below rusty buildings. Freshest (gray) dumps nearby also yield pyrite cubes up to 1½ inches across. Crystals may be found in great abundance on the dumps in this area, but some of the roads are hazardous.

Continuing log:

2.5 Turn right on Colo. 279 (not marked).

3.2 Turn left onto forked road and take right fork. Magnetite on this hillslope. Road beyond is bumpy.

3.8 Gloryhole. The rim of the pit is very dangerous and should not be approached too closely; keep children away. Jeep tours from Central City drive around the rim.

The crystals of clear, white quartz occur in abundance as coatings on ore minerals in vugs 2-3 inches across. Scepter quartz, mostly about 1 inch in size, is especially attractive. The pyrite is associated with sphalerite, galena, and quartz in one type of ore,

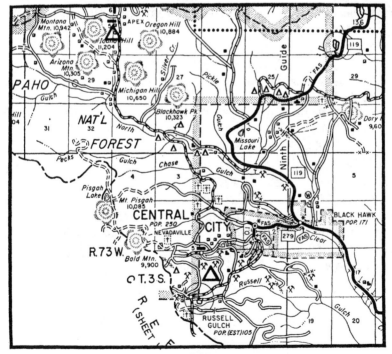

207

and with chalcopyrite, quartz, and tennantite in another. Dave Hoover suggested that the quartz and pyrite can best be seen in reflected sunlight after a rain.

The term Gloryhole refers to the opening of the San Juan mine, being named for the type of mining — characteristic also of certain African diamond mines — that combines deep open-pit and underground operations. The San Juan mine is developed by a shaft 916 feet deep, having 11 levels, which adjoin the La Crosse tunnel. The Patch itself is a partly mineralized neck of broken rock (explosion breccia) situated on the combined California-Mammoth lode west of the center of mineralization that has furnished most of the ores of the Central City area. It descends steeply from the surface about 1,600 feet, extending beneath the famous Argo tunnel, which goes between Central City and Idaho Springs, and crosses the Patch for a distance of 545 feet. The property has been worked intermittently since 1888, the main period of activity being from 1890 to 1894, and has produced about $600,000 in gold, silver, and copper.

The geology of this deposit was described by Edson S. Bastin and James M. Hill (Ref. 1). Additional information was furnished by T. S. Lovering and E. N. Goddard (Ref. 2).

REFERENCES

1. U. S. Geological Survey Professional Paper 94, 1917, p. 96-97, 233-237.
2. U. S. Geological Survey Professional Paper 223, 1950, p. 84, 171-172, 181, 183.
3. U. S. Geological Survey Geologic Quadrangle Map CQ 267, 1964.

MAPS

Topographic maps: Central City (two scales; 1903-10, 1942), Denver Mountain Area (1938-48; contour or shaded relief edition, $1.00).
National forests: Arapaho, Pike, Roosevelt.

BOULDER AREA

The great tungsten district of Colorado, once the most productive in the United States, begins at Arkansas Mountain, about 4 miles west of Boulder, and continues westward at intervals for 9½ miles to the vicinity of Nederland. Specimens of brilliant, black ferberite may be found at a number of abandoned mines in Boulder Canyon. The Oregon mine, in Gordon Gulch, is located as described below. A detailed description of the district, together with an annotated bibliography of previous publications, was given by T. S. Lovering and Ogden Tweto (Ref. 1).

The Flatirons Mineral Club meets at the Home Savings and Loan Association, 1913 Broadway, in Boulder, on the fourth Friday of each month. Visitors are welcome.

The Henderson Museum of the University of Colorado contains many specimens of interest to gem and mineral collectors. It is open daily from 8:00 to 5:00 and on Sunday from 2:00 to 5:00.

Mineral collecting at Gold Hill, also in Boulder County, has been described by Richmond E. Myers (Ref. 2). He deals mostly with the sulfide minerals found on mine dumps and with the Copper King nickel mine, which yields uncommon minerals of little specimen value.

Radioactive cerite, a rare mineral found at only five places in the world, occurs at Jamestown, a few miles north of Gold Hill, but this property is not accessible by ordinary automobile; instructions for reaching it otherwise may be obtained from Everett Walker at the general store in Jamestown, or by other local inquiry. Instead, the collector may secure massive, purple fluorite, which fluoresces under long-wave ultraviolet light, on the mine dumps of the General Chemical Division of the Allied Chemical Corporation, at the office of which permission may be obtained in either Boulder or Jamestown.

Other important mining districts in the Boulder region include Ward, Magnolia, Eldora, and Caribou-Grand Island. Old silver and lead workings of the Caribou mine, at Caribou, yielded some pitchblende during the uranium boom. These districts were described by Thomas S. Lovering and Edwin N. Goddard (Ref. 3).

REFERENCES

1. U. S. Geological Survey Professional Paper 245, 1953.
2. Rocks and Minerals, vol. 22, 1947, p. 203-206.
3. U. S. Geological Survey Professional Paper 223, 1950.

MAPS

Topographic maps: Boulder (two scales, 1957), Denver Mountain Area (1938-48; contour or shaded-relief edition, $1.00), Eldorado Springs (1965), Nederland (1942), Tungsten (1942), Ward (1957). National forests: Arapaho, Roosevelt.

Oregon Mine

The Oregon mine is the best collecting locality in the Boulder area, described above, according to J. E. Byron, Boulder mining engineer and former U. S. Mineral Surveyor. This mine is the largest in a group formerly owned by the Vanadium Corporation of America but no longer in operation. It is now owned by Cold Spring Tungsten, Inc. Specimens are readily available on the dumps, although the underground workings do not appear safe.

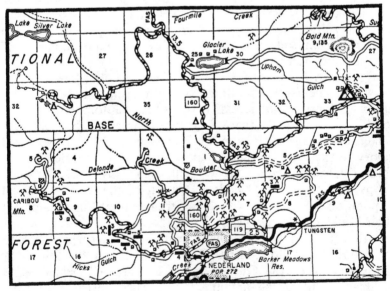

The ferberite occurs as black veinlets and crusts; the small, shiny crystals are found in vugs and on open surfaces; do not mistake the common biotite mica, which can be flaked with a knife, for ferberite. Purple fluorite and common sulfide minerals are among minerals seen here. The mine was described by T. S. Lovering and Ogden Tweto (Ref. 1).

This mine can be reached from Nederland by the following log:

0.0 Junction of Colo. 160 and Colo. 119. Go on Colo. 160 toward Ward.

3.7 Turn right onto Sugar Loaf-Sunset road where highway curves left.

7.0 Stop on right. Oregon mine on left across road.

REFERENCES

1. U. S. Geological Survey Professional Paper 245, 1953, p. 157-159.

MAPS

Topographic maps: Gold Hill (1957).
National forests: Arapaho, Roosevelt.

The highway from Denver to the Wyoming line follows the foothills of the Front Range through a well-settled area in Denver, Boulder, and Larimer Counties. The Owl Canyon alabaster local-

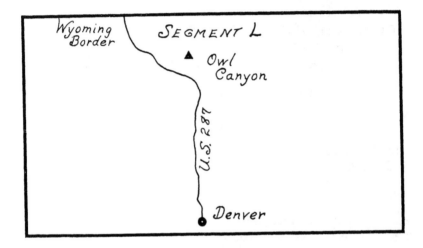

ity, described below, stands near this road.

The Fort Collins Rockhounds, Inc., meets in the Industrial Arts Building at Colorado State University (211 West Laurel Street), on the third Monday of each month except December.

The Weld County Rock and Mineral Society, Inc., meets in Greeley at the Greeley Youth Center Building, on the first and third Thursday of each month from September to May.

Owl Canyon

The largest alabaster quarry in the United States is situated in Owl Canyon, in eastern Larimer County, north of Fort Collins. Formerly operated by the Pioneer Alabaster Company, of Fort Collins, this deposit consists of beds of alabaster 3 to 4 feet thick, in the Lykins Formation, covered by the same amount of overburden. Abundant specimens of alabaster can be picked up in Owl

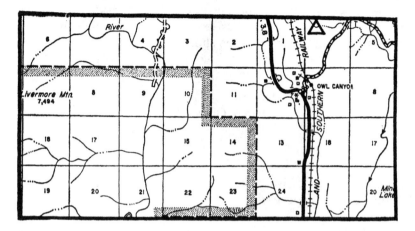

Canyon in the vicinity of this quarry and others, and attractive, rough pieces can be bought locally for a few cents a pound. Pink and white satin spar and selenite, two other varieties of gypsum, occur with the gray, brown, and mottled alabaster. The pseudo-hexagonal, twinned crystals of aragonite known as "Indian dollars" and "pioneer dollars" are found in the same formation east of Owl Canyon, according to A. J. Gude III, as reported by Edwin B. Eckel (Ref. 1). Onyx marble also occurs in this locality.

The log to this quarry from the Owl Canyon Trading Post, just off U. S. 287, 17 miles north of Fort Collins, is as follows:

0.0 Texaco station and Owl Canyon store on right. Turn right before building, cross old road onto gravel. Keep to right at first fork after cattle guard.

0.3 Keep on main road at crest of hill before crossing railroad tracks. This road joins U. S. 87 in 11 miles. (Beyond railroad tracks is abandoned alabaster quarry on left.)

1.3 Turn left through gate, and close gate. Cross pasture on dirt road toward red hogback.

1.7 Park and walk several hundred yards up gully to quarry.

From a commercial standpoint, alabaster is one of the Colorado ornamental minerals best known to outsiders. Articles fashioned from it are shipped across the country and have been sold in foreign lands. The choicest Colorado alabaster has pinkish veining and mottling, eminently suitable for making lamps, lighthouses, bookends, and ash trays that are so familiar in tourist stores. The

213

ease with which this compact variety of gypsum can be carved with a knife or on a lathe has led to its being widely used for amateur work and in small home shops.

REFERENCES

1. U. S. Geological Survey Bulletin 1114, 1961, p. 57.
2. Colorado School of Mines Quarterly, vol. 44, no. 2, 1949, p. 231.

MAPS

Topographic maps: Livermore (two scales; 1907, 1960).
National forests: Roosevelt.

Trail Ridge Road, which links both ends of U. S. 34, crosses Rocky Mountain National Park, mostly above timberline, in one of America's finest mountain highways. It passes for 4 miles along the crest of the Front Range at altitudes over 12,000 feet, the highest continuous automobile road in America. This splendid highway is usually open from about May 30 to October 25. The village of Estes Park is its famous eastern gateway, and its western entrance is at beautiful Grand Lake. Covering 405 square miles and culminating in mighty Longs Peak (14,255 feet), Rocky Mountain Park is a vivid, outdoor textbook of glacial geology, and the Specimen Mountain locality, described below, is itself a museum of ancient, volcanic action, astride the Continental Divide, north of the main road.

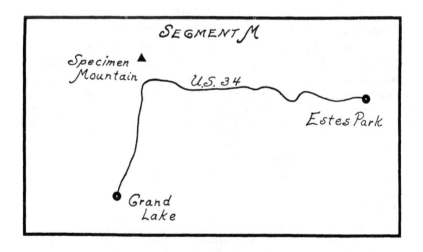

Specimen Mountain

Although collecting is not permitted in Rocky Mountain National Park, Specimen Mountain lives up to its name as a source of jasper, opal, agate, and onyx suitable for cutting.

Rising to 12,482 feet, Specimen Mountain is exactly on the Continental Divide in the northwest corner of the park, in Larimer and Grand Counties. It is reached by driving on Trail Ridge Road 31.1 miles from Estes Park to Poudre Lakes at Milner Pass (10,759 feet). A few hundred yards before the two Poudre Lakes are reached, a marked trail leads to the right into the woods. Keeping left at each fork, one reaches Specimen Mountain in about 1½ miles. The mountain is deeply glaciated, and the northwest and south sides have steep drops of as much as 3,000 feet.

David M. Seaman has described collecting at Specimen Mountain (Ref. 1). The mineralogy and geology of the area were described by Ernest E. Wahlstrom (Ref. 2-3).

Specimen Mountain is the remanant of a huge explosive volcano. Its west and south sides are flanked by various explosive and flow rocks; these contain the minerals of interest, and so specimens are confined to this part of the area. Two layers of pitchstone, a glassy rock related to obsidian, are the main mineral carriers; they average about 15 feet in thickness and are seen near the top of the productive zone.

Opal and jasper occur as a replacement in large amounts in the pitchstone. They also fill fault openings in the pitchstone, as well as in the adjacent layers of other rock. Some of the opal fills cavities in the nodules of jasper. Geodes containing opal, agate, onyx, and green chalcedony are very common. A little of the opal is transparent and almost gemmy; this and the opaque, white or yellowish opal can often be cut into attractive specimens. Material suitable for cameos has been described by Anna A. Conn (Ref. 4). The agate is green and gray in color and occurs in thin pieces, grading into thicker masses of a curious type of onyx, which has alternating layers of opal and chalcedony. Other minerals that have been found in the geodes include nontronite (formerly called chloropal), allophane, and platy calcite (partly replaced by opal and chalcedony). Colorless to light-pink topaz fills some of the cavities in the rock of Specimen Mountain, which is quartz trachyte. The topaz is accompanied by tridymite and rock crystal, all three minerals being small but of fine quality.

REFERENCES

1. Oregon Mineralogist, vol. 2, 1934, p. 17.
2. Geological Society of America Bulletin, vol. 55, 1944, p. 77-89.

3. American Mineralogist, vol. 26, 1941, p. 551-561.
4. Pennsylvania Academy of Science Proceedings, vol. 13, 1939; p. 134-135.
5. U. S. Geological Society Bulletin 1114, 1961, p. 25, 39, 139-140, 243, 246, 278, 334, 339.

MAPS

Topographic maps: Denver Mountain Area (1938-48; contour or shaded-relief edition, $1.00), Fall River Pass (1958), Rocky Mountain National Park (1961, contour or shaded-relief editions, $1.00).
National forests: Arapaho, Roosevelt.

The highway east from Denver to Sterling – through Denver, Adams, Weld, Morgan, Washington, and Logan Counties – crosses the Great Plains on the way to the Kansas border. It passes the roads that lead to the good jasper locality and the famed blue-barite locality near Stoneham, both described below.

Occupying the eastern two-fifths of Colorado, the Great Plains are on the whole less spectacular than other parts of the state, but they have an attractiveness of their own, though a different kind. Scenically, they are dramatic contrast to the grandeur of the mountains. Historically, they are rich in interest, presenting a pageant of greatness that reaches down the centuries from Indian to frontiersman to pioneer.

"Grass, sunshine, and solitude" characterize the High Plains division of the Great Plains, which forms a fringe around the northern and eastern parts of Colorado, extending in places farther into the interior. Having few stream valleys and wide areas between them, the High Plains constitute an extremely flat region mantled with sand, gravel, and silt brought down from the Rocky Mountains on the west. West of the High Plains is the Colorado Piedmont, which has much greater differences in elevation, indi-

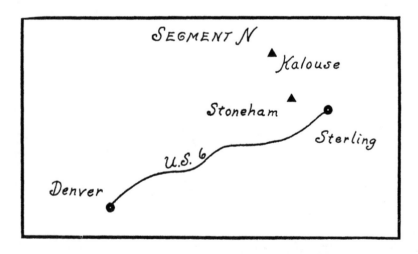

cating the effect of streams near the mountains in removing a part of the material that had previously been deposited, leaving mesas, buttes, and other landforms on an irregular surface. Pawnee Buttes, a world-famous locality for vertebrate fossils of Tertiary age, is described in *Exploring Rocks, Minerals, Fossils in Colorado*, by Richard M. Pearl (Sage Books, The Swallow Press, Inc., Chicago, revised edition, 1969).

Rev. John F. Stein, of Iliff, has described "Rockhounding Along the South Platte River" (Ref. 1), in which he names many interesting finds made by him and his students. These include petrified wood, blue agate, red agate, white-lace agate, and petrified bones of various prehistoric animals, as well as miscellaneous chalcedony and the Stoneham barite.

In Akron, 34 miles south of Sterling, the High Plains Rock and Artifact Society meets on the third Monday of the month at the Y-W Electric Building.

REFERENCES

1. Lapidary Journal, vol. 10, 1956, p. 34-36.

MAPS

The Index to Topographic Maps of Colorado, obtainable free from the U. S. Geological Survey, Federal Center, Denver, Colorado 80225, shows which parts of northeastern Colorado are covered by topographic maps.

Kalouse

Colorado's finest jasper is found associated with agate and petrified wood near the former settlement of Kalouse, in northeastern Weld County. Kalouse, marked by a farmhouse (with gas pump) on the southwest intersection corner, is situated 14 miles north of New Raymer, 37 miles north of Fort Morgan, and about 10 miles southeast of Pawnee Buttes. It is most readily reached by going north (on a gravel road) from Colo. 14, from the east side of New Raymer, which is 9 miles west of Stoneham.

Specimens are found scattered in all directions from the right-angled road intersection at Kalouse. Especially favored places are along the east-west road. Some of the material in this and nearby localities resembles the renowned Fairburn agates of South Dakota

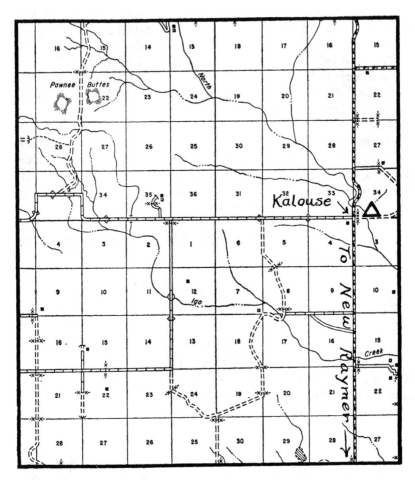

and is likewise found in Tertiary rocks, although it has weathered out onto the surface and may be picked up easily. Brown jasper of remarkable uniformity is found here in abundance.

REFERENCES

1. U. S. Geological Survey Bulletin 1114, 1961, p. 280.

Stoneham

One of the best barite localities in the United States is situated about 5 miles northeast of Stoneham, in eastern Weld County. It has long been referred to as Sterling but is actually about 26 miles from that city. The barite from here is esteemed for its exquisite, blue color, its fine transparency, and its sharply formed crystals. Ease of collecting is another virtue of this locality, although permission to collect should be obtained at the nearest farmhouse, and it is required that any holes dug must be filled before leaving. "Ease of collecting" does not imply that muscular exertion will not help to produce specimens from the tough clay.

The log to this locality from the highway intersection near Stoneham is as follows:

0.0 Intersection of Colo. 14 and Colo. 71, 9 miles east of New Raymer. Go east on Colo. 14 toward Sterling.

0.6 Pass road to Stoneham on right (unmarked in 1971).

1.1 Turn left onto gravel road (RD 149) at crest of hill.

4.2 Cross cattle guard at fork and stay right.

5.2 Turn right on faint dirt road toward bluffs. Park at edge of bluffs above badland gullies.

Crystals and fragments of barite lie on the surface and somewhat embedded in the loose rock along the sides and bottoms of

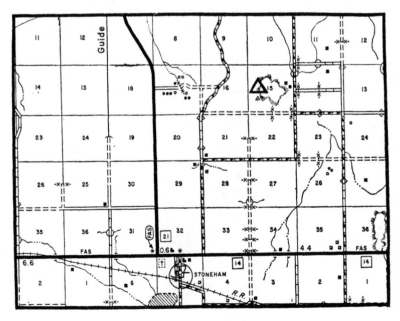

the gullies, where they have weathered out of a shale bed. The original, narrow calcite veins in the shale are at various angles and run discontinuously in every direction. They naturally yield the best unbroken material, although they are not too easily seen at a glance. With a pick, groups of crystals can be removed from these veins and also from crevices and hollows in the walls of the bluffs, where they favor horizontal veins.

These bluffs are composed of a limy clay and are known locally as chalk cliffs because of their whiteness, although it is tinged by the yellow of iron. The clay is an alteration of volcanic ash carried from the west during the Tertiary Period and buried beneath later, coarse sediment from the mountains. Recent erosion has produced the present badlands topography by stripping off the covering and exposing the clay with its layer of barite-bearing shale. Boulders lie strewn on the ground in abundance.

The choicest crystals are perfectly terminated (some at both ends), entirely transparent, and as blue as typical aquamarine. In fact, they look much like aquamarine except for their flattened crystals, but of course they are much softer. They range in length up to 6 inches, although the average is perhaps 1 to 2 inches. Clusters of the barite-rose type are found occasionally, mostly on the upper slopes of the bluffs, a short distance below the top of the escarpment, as described by Guy B. Ellermeier (Ref. 1). Opal, which fluoresces green, perhaps owing to the radioactivity that may have colored the barite blue, coats some of the specimens. Delicate crystals of golden barite also occur in certain spots.

REFERENCES

1. Rocks and Minerals, vol. 23, 1948, p. 21.
2. Mineralogist, vol. 20, 1952, p. 364.
3. U. S. Geological Survey Bulletin 1114, 1961, p. 28, 65-66.
4. Rocks and Minerals, vol. 40, 1965, p. 538-539.